# HETI Membership Directory

AF887501

# 2015/16

ISBN 978-0-9973368-0-1

# THE FEDERATION OF HORSES IN EDUCATION AND THERAPY INTERNATIONAL A.I.S.B.L




**SECRETARIAT**

HETI

P.O. Box 134 • Damariscotta, ME 04543

Email: office@HETIfederation.org • www.HETIfederation.org • www.facebook.com/HETI.international





ISBN 978-0-9973368-0-1

# HETI Membership Directory

# 2015/16

ISBN 978-0-9973368-0-1

# Table of Contents


| | | |
|---|---|---|
| 5 | - | Letter from the President |
| 6 | - | Letter from the Executive Director |
| 7 | - | Our Sponsors |
| 8 | - | HETI Level I Instructor and Certification Program |
| 10 | - | HETI History |
| 12 | - | The Mary Selway-Swift Development Fund |
| 13 | - | Our Advertisers |
| 25 | - | HETI Membership Directory 2015/16: Federation Members |
| 81 | - | HETI Membership Directory 2015/16: Business Members |
| 85 | - | HETI Membership Directory 2015/16: Associate, Institute & Honorary members |
| 91 | - | Index of Members by Country |

# Letter from the President

Dear HETI members,

I invite you to explore the content of this Membership Directory. For me it is always interesting to see who our members are all around the world and what they do in the field of EAA/EAT.

Inside this directory there you'll get a glimpse of the wonderful cultural characteristics of every association. I find it very interesting reading and I hope you do too. It is like a trip around the world, seeing pictures and reading about the activities and educational offerings given in different countries. Learning about them gives one a perspective on one's own work and could perhaps inspire you to do something new in your own organisation!

The common denominator here is of course the horses which play the leading role in our work, in addition to your work with people with special needs. The superb quality of your work, both with clients and in your organisations, and your ongoing commitment, can be read "between the lines" in the descriptions contained inside this book. I´m proud of your work!

The Executive Board elected in Taiwan in the XV International Congress has been meeting via Skype and using documents in "the cloud." EC members living in different time zones have now a reasonable opportunity to work on a task in their own time. We have made a strategic plan for 2015-2018, which is an important task; we are outlining ethical principles for horse use, stabling and equine welfare in therapy which I hope you all can agree upon when they are published.

At this time I have the opportunity to announce the awarding of HETI level I education certificates in Greece, Taiwan and Japan. Congratulations to you! You can contact Gisela in the office to obtain more information on this valuable program.

There is strength in unity. I´m honoured to serve as your president,

Sanna Mattila-Rautiainen,

President, HETI

# Letter from the Executive Director

Dear HETI Members:

I am happy to write you again on the occasion of the publication of our 2015-2016 HETI Membership Directory. I think of this as a portrait of our organization that consists of 247 members in 48 countries.

This book is a resource for you to use and to refer to often. I hope you will use it, and even more importantly, share it with others. It is not only a snapshot of HETI, but a means of connection with others, which is the only way HETI can thrive and grow.

It has been a fruitful and productive year for HETI with many positive and exciting developments coming on the heels of the XV International Congress in June which was superbly produced by the Taiwan organizing Committee chaired by Uta Rindfleisch, the Department of Sports Management of Aletheia University, and the members of the two local Federation members, the Chinese Taipei Equestrian Association and Qi Zhi Vocational Training Centre. At the Congress we elected a new president, Sanna Mattila Rautiainen, as well as an energetic and inspired Executive Board.

Our revamped membership structure has yielded many new members. We are very pleased to welcome new Federation members from the Czech Republic, Ireland and Wales as well as new Associate, Business and Institute members. The result of this is a more diverse and varied membership roster that strives to offer value to every type of constituent. Our newsletter, website and our peer-reviewed Scientific and Educational Journal of Therapeutic Riding remain major sources of information for our members and those who are interested in equine assisted activities. I encourage you to submit scientific papers to our editors for review at editorial@hetifederation.org and for the newsletter at office@hetifederation.org.

We look forward with excitement to our next Congress, two years hence, to be held in Ireland under the capable leadership of Jill Carey and Festina Lente. Thanks to so many of you who keep me informed during the year. Your thoughts, comments and dreams are always welcome. Please stay in touch with me via email, the phone and of course our Facebook page.

With best wishes,

Gisela H. Rhodes, Executive Director/ Secretary, HETI

# Our Sponsors

**BETZ HAARTZ, USA, BRONZE (IN MEMORY OF VERNON RHODES)**
**PETALS, INC., AUSTRALIA, BRONZE**

# Become a Sponsor!

HETI is at the heart of international Equine Assisted Therapy Education. Your sponsorship makes it possible!

- Support our members in spreading cutting edge research results through Newsletters, Publications and International Congresses.
- Help us promote the field of Equine Assisted Activities through literature, lectures and hands-on workshops.
- Support humane horsemanship methods to ensure that our partner, the equine, benefits and enjoys activities just as much as the humans.

Of course, contributions in any amount are appreciated. Interested? Please contact Gisela Rhodes at office@HETIfederation.org. Sponsorship benefits include:

### PLATINUM SPONSOR: $25,000 AND UP

An all-expenses-covered invitation to an International Congress
Invitation to meet the President and the Executive Committee of HETI and attend all functions as the special guest of the President
Plus, all lower tier benefits

### BRONZE SPONSOR: $200 - $999

Listing in HETI's Annual Membership Directory
Listing in frequent e-newsletter
Logo and link on HETI's website

### GOLD SPONSOR: $10,000 - $24,999

Invitation to visit a HETI member program in the country of your choice
Attend a personal tour of the program
Plus, all lower tier benefits

### SILVER SPONSOR: $1,000 - $9,999

Establish a special scholarship fund in your name
Display your logo at the HETI booth during the International Congress
Plus, all lower tier benefits

# HETI Level I Instructor and Certification Program

The Horses in Education and Therapy International (HETI) Level I Instructor Training and Certification Program provides the basic educational back-ground to facilitate the certification of instructors who will be able to provide knowledgeable, progressive and safe equine experiences for students who are physically, cognitively, or emotionally disabled. HETI Level I Instructor and Certification Program

Level I Instructor Training and Certification is comprised of three phases. Phases I and II include theoretical and practical learning in the areas of:

- design and management of the riding program
- building the team
- the horse
- definition of physical, cognitive and emotional disabilities
- developmental and postural responses
- tack and adaptive equipment
- horse handling
- equine gait analysis
- developing your eye
- horse and rider working together
- selection and training of mounts
- methods of knowledge transfer
- therapeutic riding techniques
- mounting and dismounting techniques
- use of volunteers and assistant instructors
- goal setting, lesson planning
- hands-on teaching

Phase III is the certification phase.

***For more information, contact the HETI Office at office@hetifederation.org***

# HETI Instructor Code of Ethics and Conduct

Competence and Teamwork are the foundation of practice for a HETI Therapeutic Riding Instructor. HETI expects its instructors to conduct themselves in an ethical and professional manner to ensure the integrity of the organization, riders, volunteers and equines.

HETI Instructors are committed to providing a safe environment for all participants and horses and to using riding and training methods to achieve harmony and understanding between horses and riders. They champion the humane treatment and well-being of horses and respect the diversity of all riders and horses.

They use best practices and standards which are based upon generally accepted methods. Each country and culture may adapt these best practices to their own environment; however, adaptations should always consider the well-being and safety of our riders and our horses.

# HETI Level I Certified Instructors

Congratulations to the following people who completed their HETI certification in Taiwan and Greece. Another certification program has been started in Indonesia and Portugal.

- Chao-Yuan Chang (Taiwan)
- Chun-Mei Chen (Taiwan)
- Jose Montilla (Phillipines)
- Yi-Hung Chang (Taiwan)
- Ya-Hui Li (Taiwan)
- Shu-Ya Chan (Taiwan)
- Tsung-Hsin Weng (Taiwan)
- Yu-Shan Lu (Taiwan)
- Sara Hutton-Anetopoulou (Greece)
- Lappa Gkampriella (Greece)
- Nantina Panagou (Greece)
- Tatjana Richter (Greece)
- Ekaterini Chomoriti (Greece)
- Anna Mitropoulou (Greece)
- Tze-Hsuan Wang (Taiwan)
- Ming-Hui Chang (Ella Chang - Taiwan)
- Pei-Ying Lee (Koala Lee - Taiwan)
- Yu-Ling Tsai (Jade Tsai - Taiwan)
- Tata Montilla (Philippines)

# HETI History

**Past HETI Presidents:**

Dorothee Ames, Canada;
Dr. Carl Klüwer, Germany;
Gillian Lawrence, Canada;
Octavia Brown, USA;
Gundula Hauser, Austria;
Pauline Perry, Australia;
Ann O'Shallie, USA

An International Congress is held every three years and offers an opportunity for all of us to get together, exchange ideas and celebrate our achievements in the world of equine assisted activities. Our past congresses are:

**1974 I. International Congress in Paris, France**
*Reeducation par L'Equitation*

**1976 II. International Congress in Basel, Switzerland**
*Rehabilitation through Riding - Riding Therapy*

**1979 III. International Congress in Warwick, United Kingdom**
*Riding as a Sport for the Disabled*

**1982 IV. International Congress for Therapeutic Riding in Hamburg, Germany**
*Overview of the international situation in Therapeutic Riding*

**1985 V. International Congress in Milan, Italy**
*Rehabilitation through Riding*

**1988 VI. International Congress for Therapeutic Riding in Toronto, Canada**
*Therapeutic Riding and Scientific Research*

**1991 VII. International Congress for Th. R. on Aarhus, Denmark**
*An Update on Therapeutic Riding*

**1994 VIII. International Congress in Hamilton, New Zealand**

# HETI History (continued)

**1997 IX. International Congress in Denver, USA**
*Riding the Winds of Progress*

**2000 X. International Congress in Angers, France**
*Cheval et Différences*

**2003 XI. International Congress in Budapest, Hungary**
*The Complex Influence of Therapeutic Riding*

**2006 XII. International Congress in Brasilia, Brazil**
*Contact between Friends*

**2009 XIII. International Congress in Münster, Germany**
*Horses for Body, Mind and Soul*

**2012 XIV. International Congress in Athens, Greece**
*Horses in Education and Therapy*

**2015 XV. International Congress in Tapei, Taiwan**
*Horses and People*

## Our next Congress will be:

## 2018 - XVI. International Congress in Ireland

**June 25-29, 2018, at**
**Royal Dublin Society,**
**Dublin, IRELAND**

# Apply to the Mary Selway-Swift Development Fund!

Mary Selway-Swift

This fund was established from a bequest to HETI from the estate of the late Mary Selway-Swift.

Mary was involved in the field of therapeutic riding over many years - in the United Kingdom and in South-East Asia (specifically in Hong Kong and Singapore). It was Mary's wish that this fund be used to promote and improve the educational opportunities, training and quality of riding therapy through the development of therapeutic riding programs internationally. She was particularly concerned for those in developing countries or in places which have limited access to training opportunities. Any application needs to be supported by a HETI Full Member and payments from the fund will be made on the recommendation of the HETI Executive Committee.

For an application please contact the office at:

HETI
P.O. Box 134
Damariscotta, ME 04553, USA
Email: office@HETIfederation.org

Members are encouraged to promote the Mary Selway-Swift Development Fund among friends and interested people/businesses. Donations to the fund should be sent to the Office.

# Thank you to our Advertisers

This directory is partially funded by our advertisers. We would like to extend our heartfelt gratitude to the following for supporting this publication:

- Associazione Lapo
- Carlisle Academy
- Equest
- Festina Lente
- FN Verlag
- HHRF
- PATH International
- Special Equestrians (Fitting In)
- Special Equestrians (Hello Bob)

Their advertisements appear on the pages that follow.

The HETI Directory is distributed to all members, advertisers, and sponsors in forty-four countries; it's our calling card. Its reach extends to almost anyone who's influential in the field of equine-assisted activities. If you would like your organization included with an advertisement, consider the upcoming edition.

For rates and placement in the 2017/18 Membership Directory please contact Gisela Rhodes at the HETI office:

P.O. Box 134
Damariscotta, ME 04543
Email: office@HETIfederation.org
www.HETIfederation.org

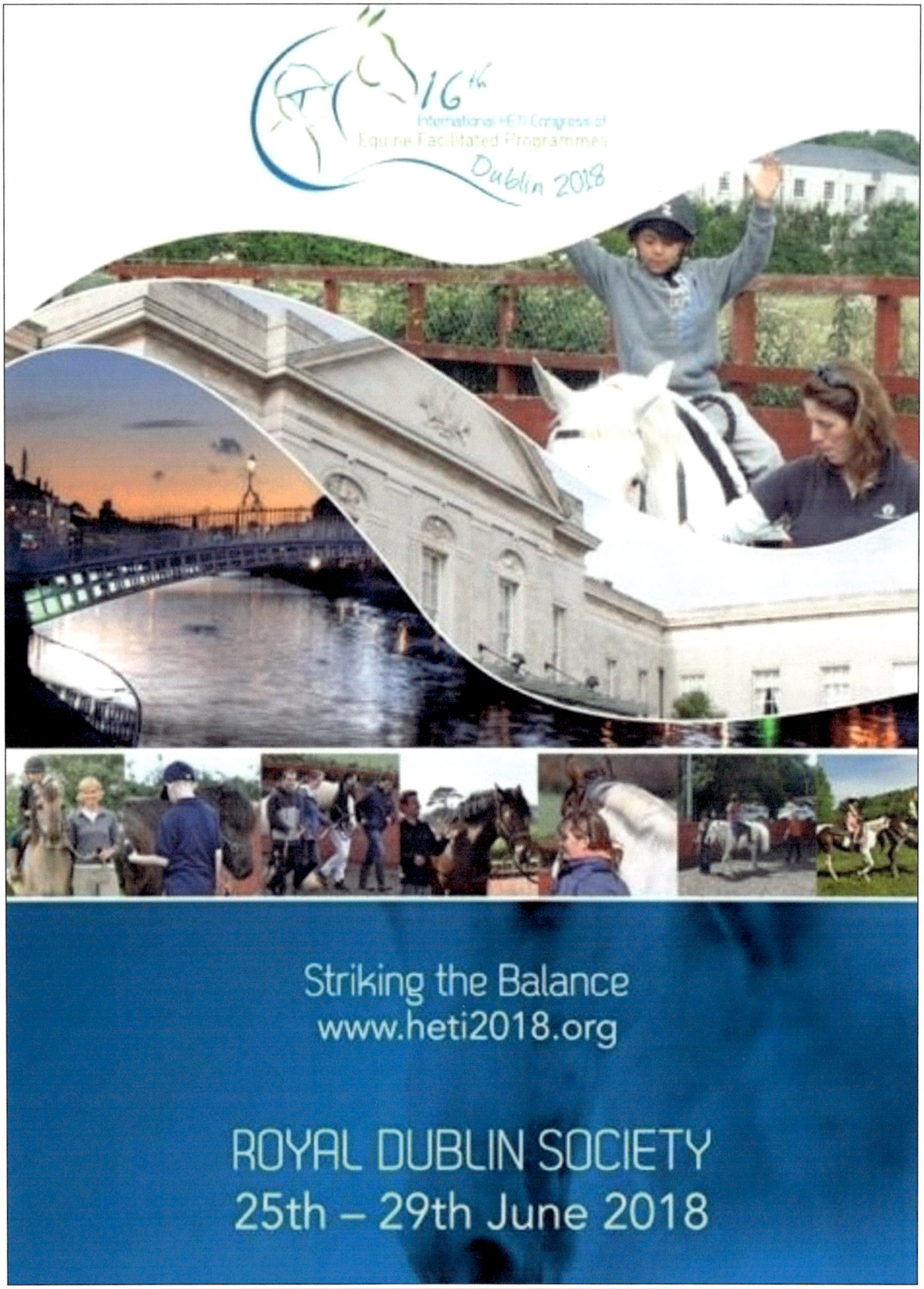
16th
International HETI Congress of
Equine Facilitated Programmes
Dublin 2018
Striking the Balance
www.heti2018.org
ROYAL DUBLIN SOCIETY
25th – 29th June 2018

# STRIKING THE BALANCE IN IRELAND AT THE HETI 2018 CONGRESS

The key note speakers have been invited based on their experience and expertise within four key pillars of Equine Assisted Programmes:

## Learning and Development – Physical and Mental

Regardless of your particular discipline, your ability to facilitate positive and rewarding experiences for young people and adults is essential. The 2018 Congress will focus on this area and will be addressed by William Micklem, who has worked extensively and successfully in illustrating strategies and techniques to help people improve their performance.

## Sustainability

Creating sustainability requires a number of factors including strategic planning, budgets, cash flow projections, marketing, human resource management (volunteers and paid staff), mission statements, governance, identification of values, asking what our clients – our customers – want and expect from us.....and more!

## Equine Welfare and Management

For horses in therapeutic riding or hippotherapy, we are mindful of rider weights and the need to ensure a balanced training and working schedule for horses, how bits can potentially cause levels of discomfort to horses and the need to train horses to habituate to a wide range of sudden movements by riders, loud noises and other unpredictable behaviours by riders.

## Research

As the 2018 congress organisers, we are keen to bring the latest EAP research to Ireland – and we hope that each delegate will leave with strong and solid research findings that will help them improve their own EAP. However, we are also mindful that we all started somewhere and that there are many EAP practitioners out there who are practising very valuable work and may be at the exploratory stages.

# PATH Intl. centers and professionals

## insist upon **safety** and focus on **possibilities**

Since 1969 the Professional Association of Therapeutic Horsemanship International (PATH Intl.), formerly the North American Riding for the Handicapped Association, has been ensuring excellence and changing lives through equine-assisted activities and therapies (EAAT), which promote health, fitness and socialization for people with special needs.

PATH Intl. member centers, certified instructors and credentialed professionals belong to a progressive and growing industry. They offer safe, ethical and humane practices in the rapidly growing field of EAAT. Nearly 62,000 participants of all ages with a variety of special needs have made the right choice by trusting their safety and growth to professionals providing opportunities in therapeutic riding, interactive vaulting, carriage driving or mental health and learning sessions at PATH Intl. centers.

Nearly 870 PATH Intl. member centers and more than 4,600 certified professionals are trusted and operating within industry best practices while implementing the top safety standards in the business.

**Learn more about equine-assisted activities and therapies, get involved, make a donation and find your local PATH Intl. center:**
**www.pathintl.org**
**(800) 369-RIDE (7433) • PO Box 33150, Denver, CO 80233**

PATH
INTERNATIONAL
Professional Association of Therapeutic Horsemanship International

*Ensuring excellence and changing lives through equine-assisted activities and therapies®*

# Scientific and Educational Journal of Therapeutic Riding

## Now Peer-Reviewed!

**The SEJThR is an international journal with the purpose of advancing the field of Equine-Assisted Activities and Therapies. Current research, case studies, and topic discussions are part of the journal content.**

- ***Are you or your center conducting a research study?***
- ***Are you using innovative approaches with a client?***
- ***Are you knowledgeable about a specific topic area in EAAT?***

Consider submitting a manuscript to the SEJThR! We also welcome additional reviewers to support the peer-review process! Go to http://hetifederation.org/journal to read abstracts, learn how to access journal articles, and for submission guidelines.

**The Federation of Horses in Education and Therapy International A.I.S.B.L.**

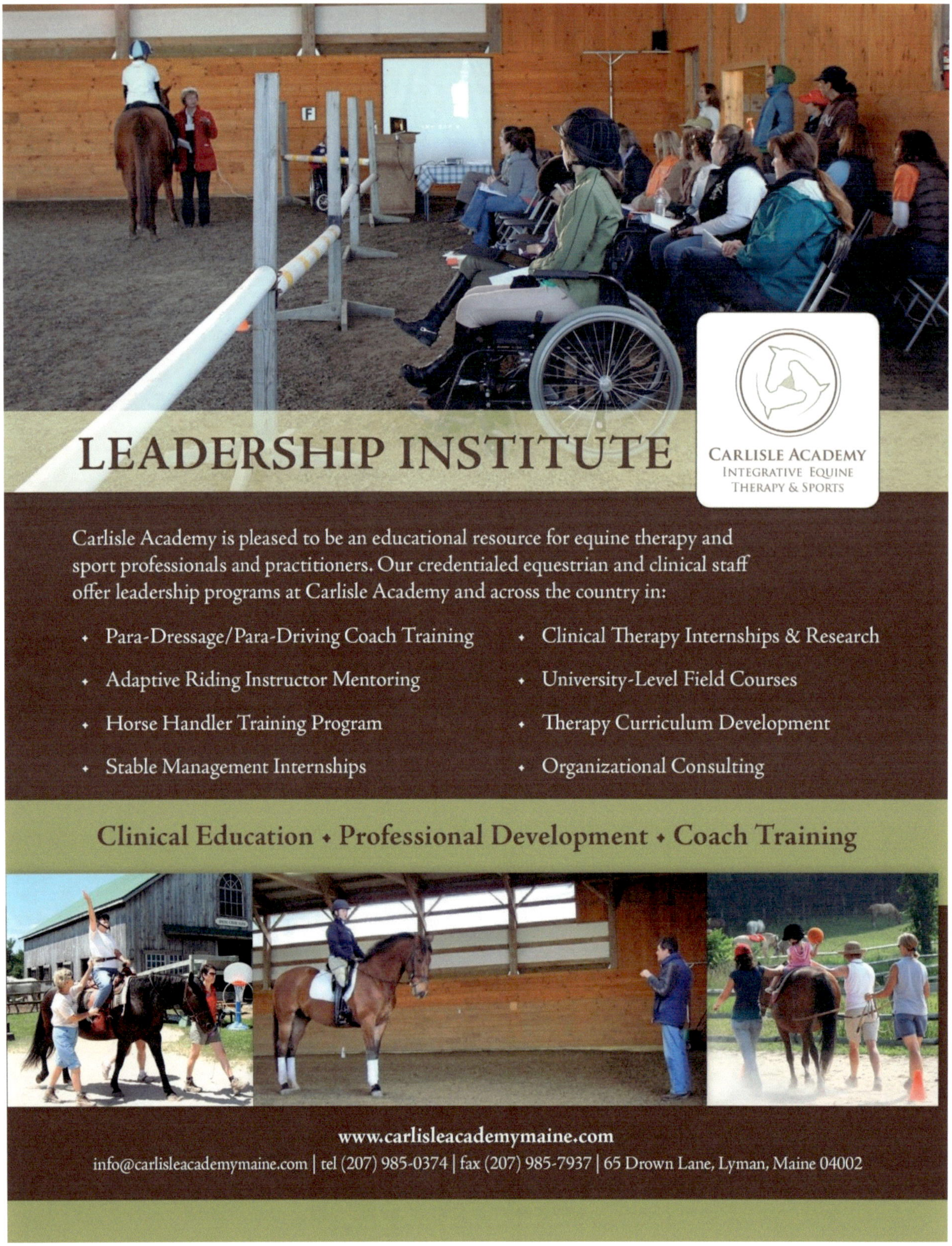
LEADERSHIP INSTITUTE
CARLISLE ACADEMY
INTEGRATIVE EQUINE
THERAPY & SPORTS
Carlisle Academy is pleased to be an educational resource for equine therapy and sport professionals and practitioners. Our credentialed equestrian and clinical staff offer leadership programs at Carlisle Academy and across the country in:
• Para-Dressage/Para-Driving Coach Training
• Adaptive Riding Instructor Mentoring
• Horse Handler Training Program
• Stable Management Internships
• Clinical Therapy Internships & Research
• University-Level Field Courses
• Therapy Curriculum Development
• Organizational Consulting
Clinical Education • Professional Development • Coach Training
www.carlisleacademymaine.com
info@carlisleacademymaine.com | tel (207) 985-0374 | fax (207) 985-7937 | 65 Drown Lane, Lyman, Maine 04002

# Have you ever tried to keep six students engaged and happy… and only had one horse?

**Hello Bob! + 49 Other Interactive Vaulting Games** is written for vaulting instructors and coaches who are looking for creative activities besides the vaulting compulsories. It is also an equally valuable tool for instructors of able-bodied and therapeutic riding sessions. In plain, often humorous language Gisela H. Rhodes shares her secrets accumulated during fourteen years of experience with therapeutic riding. Each game features specific information on skills addressed, set-up and materials, and detailed instructions on how to play the game safely for horse and student.

A gem of a book!

*"**Finally**, we have a collection of activities that Gisela has used for many years in her vaulting workshops. Some are very simple and others much more challenging - all proven to be very effective, educational and motivating. All of us at Windrush are delighted that this much anticipated book is finally available as a teaching tool. Both, students and instructors will be inspired and will benefit from this work."*

Amanda Hogan, Executive Director of Windrush Farm Therapeutic Equitation,
PATH INTL. Master Instructor

**Hello Bob! + 49 Other Interactive Vaulting Games**

**ISBN: 978-0-9801325-2-6**

**$22.95**

**Shipping:**
**$5 plus $1 per book**

Gisela H. Rhodes, M.Ed., is a PATH Intl. (NARHA) Interactive Vaulting and Advanced Therapeutic Riding Instructor, as well as a Centered Riding© Instructor Level 1. She has served as the chair of the PATH Vaulting committee, as an EFMHA Board member and is currently executive director of HETI (FRDI). Her workshops, lectures and clinics are sought after, both nationally and internationally. Born in Germany, she lives at Fair Tide Farm in Newcastle, Maine.

**Special Equestrians, Inc. • 220 River Road • Newcastle, Maine 04553 • www.fairtidefarm.com • (207) 563-3974 • info@fairtidefarm.com**

# Professional Equestrian Competence

ISBN 978-3-88542-444-4

€ 36.00

## Know How, Understand Why, Feel When

In this book Michael Putz comments on the German Guidelines Volumes 1 & 2 and presents clear evidence that serious, carefully considered training, carried out feelingly in accordance with the rules of classical equestrianism, is the system which is ideally suited to horses and most beneficial for them. His prime objective, to convey the real training of horse and rider, is achieved by means of his exceptional ability to recognise problems, to analyse them and provide solutions which are possible to implement in practice. The result of this form of training is the maintenance of the horse´s pleasure in working together with the rider, it's willingness to perform and the „personality" of the horse as well as, ultimately, a harmonious partnership between rider and horse.

ISBN 978-3-88542-748-3

€ 29.80

## Case Studies and International Reports

Description and an in-depth look at the development of the many varied techniques used with a horse in psychological therapy.
11 different authors from Germany, Finland, the Netherlands and the United States of America present their work in this book and hope to contribute to a qualified exchange and lively discussion.

ISBN 978-3-88542-488-8

€ 18.80*

## Function and Development of the Muscles of the Riding Horse

The authors explain the physiological facts which are the foundation of the teaching of riding and discuss important factors in the training of the horse and how to recognise and rectify training problems.

ISBN 978-3-88542-474-1

€ 16.80

## Equestrian Dictionary German - English - French

A competent author as well as experts and native speakers who are well-versed in equestrian terminology guarantee the correctness of the specialised expressions and relevant translations.

FNverlag
der Deutschen Reiterlichen Vereinigung GmbH
Experts in all Equestrian Matters

*Postbox 11 03 63*
*48205 Warendorf*
*Germany*

*The Publishing House of the German Equestrian Federation*

*→ Tel. +49 (0)2581 6362-154/-254*
*Fax +49 (0)2581 6362-212*
*E-Mail: vertrieb-fnverlag@fn-dokr.de*

**www.fnverlag.com**

* recommended retail price

# HIPPOTHERAPY SADDLE "ELISA 2"

"LAPO" ASSOCIATION, IN COLLABORATION WITH "PRESTIGE ITALIA" SADDLERY, HAS REVISED ITS PREVIOUS HYPPOTHERAPY SADDLE, ELISA 1.

THE NEW "HIPPOTHERAPY SADDLE ELISA 2" IS CAPABLE OF MEETING THE SPECIAL REQUIREMENTS OF SAFETY AND ABOVE ALL, A MORE CORRECT POSTURE. SEAT, FLAP, CANTLE OF "ELISA 2" ARE CONFORMED IN SUCH A WAY AS TO PROMOTE A BETTER CORRECT SEAT AND TO FACILITATE PELVIC POSITION, TRUNK-STRAIGHTENING AND BALANCE. THE SPECIAL TRAPEZOID-SHAPED EXTRACTABLE HANDLE, WITH A PARTICULAR INCLINATION, THAT ALLOWS A SECURE GRIP AND A CORRECT POSITION OF TRUNK AND SHOULDERS, IS NOW PROVIDED BY A NEW SAFETY DEVICE.

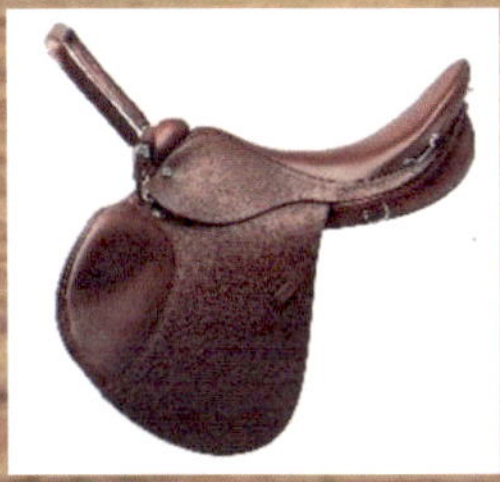

*To place an order please refer to the manufacturer, Prestige Italia: www.prestigeitaly.com info@prestigeitaly.com*

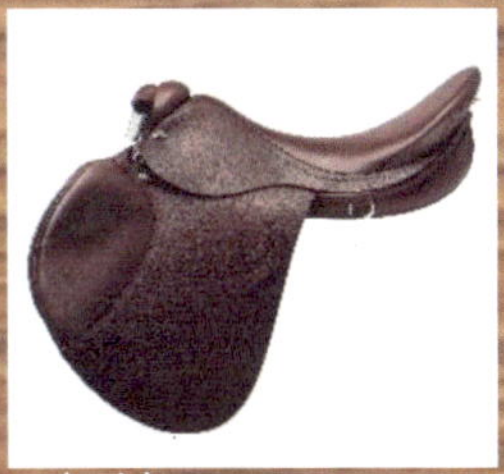

**Technical Features:**

The "Elisa 2" saddle has a medium deep seat, with round cantle. It comes with flocked panels; it is offered with a flaps' forward projection standard/ + 2cm or + 4cm, depending on the different pathologies. Elisa 2 is provided with a set of medium front and back blocks with Velcro attachments to help riders find the correct position that best suits their needs.

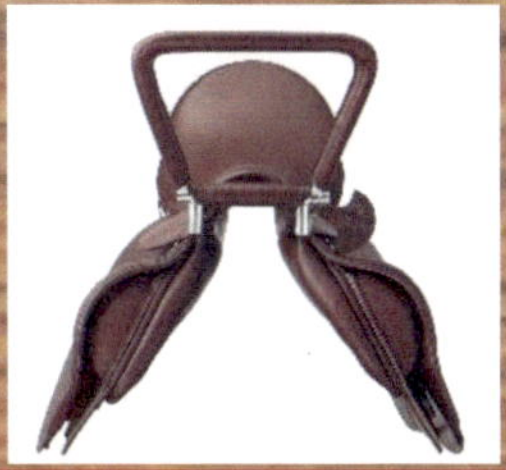

**Associazione "Lapo" ONLUS Sezione "L.R.E."** - Via A. Susini, 14 - 50125 Firenze (Italy)
Phone: +39 347 5398994 - Fax +39 055 4298432
associazionelapo@gmail.com - www.associazione-lapo.it

petals FLORIST NETWORK

# Support HETI

when you send flowers with Petals Network 20% of the flower value of your order will be donated to HETI

to order please visit:
www.petals.com.au/frda

Choose from our magnificient range of over 100 products.
There's something to suit every occassion & every budget !

# Fitting In - A Horse's Tale

This charming story about Picasso, a horse who had to face a lot of changes, is a therapeutic and educational tool for programs working with children navigating the difficult passage through adolescence.

*"What a great non-judgmental way to teach children (and adults!) about relationships, our individual differences and mutual acceptance."*

Lee Junker, Philadelphia Public Mental Health System

ISBN: 978-0-9801325-0-2 • $24.95 • plus shipping & handling

Special Equestrians, Inc. • 220 River Road • Newcastle, ME 04553, USA • www.fairtidefarm.com • +1 (207) 563-3974 • info@fairtidefarm.com

Through investment in rigorous research, Horses and Humans Research Foundation will serve as a catalyst to advance knowledge of horses and their potential impact on health and wellness of people.
www.horsesandhumans.org
Join Us
Apply for a research grant • Donate •
Be a research partner • Like us on Facebook
H&H RESEARCH FOUNDATION
Horses and Humans Research Foundation
P.O. Box 480 • Chagrin Falls, OH • 44022 • info@horsesandhumans.org

# HETI Membership Directory 2015/16

# Federation Members

## ACPTR – Association of Chartered Physiotherapists in Therapeutic Riding

Thorganby
York Y019 6DN
UNITED KINGDOM
Contact: Elizabeth Beckerlegge,
Phone: +44 (0) 190 444 8755
Email: elizabeth@beckerlegge.com

ACPTR is a professional clinical interest group of the Chartered Society of Physiotherapists in the United Kingdom. Membership of ACPTR is not restricted only to Physiotherapists.

The function of ACPTR is to promote hippotherapy (horse assisted physiotherapy treatment) and best practice by physiotherapists working within a therapeutic riding environment.

ACPTR provides a post graduate course for therapists who choose to use the horse and its movement therapeutically and for Physiotherapists to qualify in Hippotherapy. It also organises workshops, study days and a biennial conference.

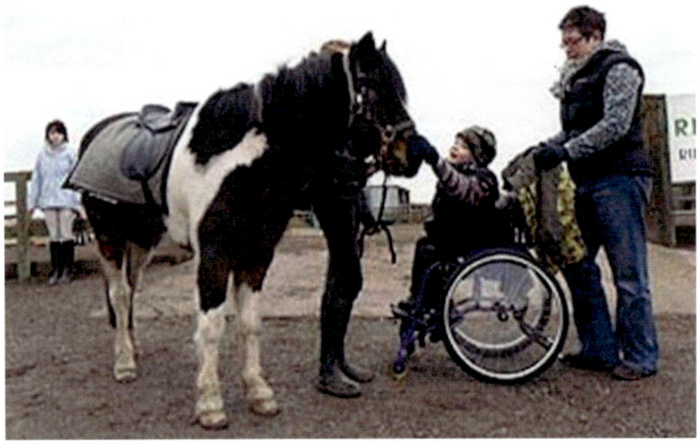

## AEDEQ – Asociación Española de Equinoterapias

Calle Calderón 71, 2º, 5ª
Sabadell 08201
SPAIN
Contact: Lorenzo Lucarelli
Phone: 00 34 686331750
Email: lucarellil@hotmail.com
Website: www.aedeq.org

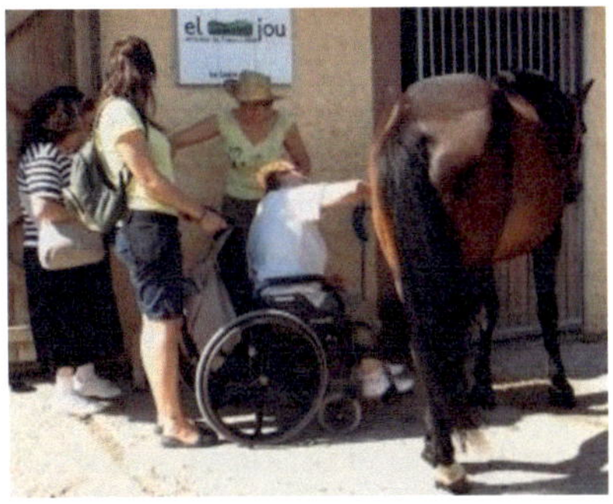

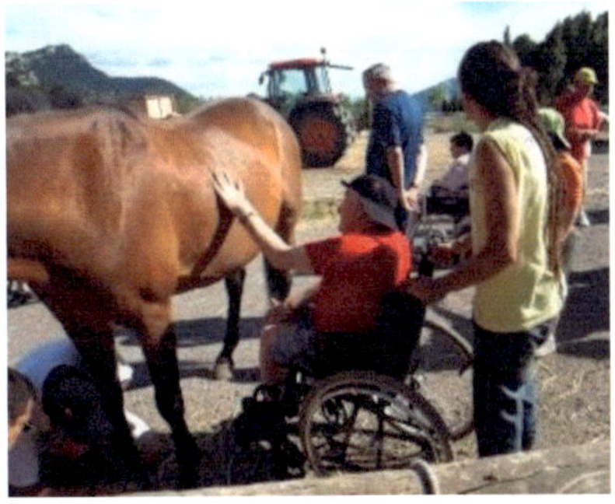

AEDEQ, the Spanish association for Therapeutic Riding has provided hippotherapy and equine assisted psychotherapy for several years and recently moved to a new stable. Since the organization's inception, it has been active in therapeutic riding as well as in leisure, sports and educational activities assisted by horses. The association is now in the process of organizing new services and incorporating them into their offerings:

- research the legal, economic, ethic, logistic, and social aspects of equine therapies
- organize various courses for therapists and instructors, at all levels, from initial to master
- incorporate therapeutic driving and vaulting
- prepare riders for adapted equestrian sports

AEDEQ is also active in promoting the REDEQUI cooperation project which is an international network of pedagogic centers providing therapeutic riding. The REDEQUI cooperation project uses tourism as a source of income to support the therapeutic team and to extend accessibility to therapeutic riding to those who cannot afford it otherwise.

## AHA – American Hippotherapy Association

PO Box 2014
Fort Collins, CO 80524
USA
Contact: Jacqueline Tiley, Executive Director
Phone: (570) 224 4505
Email: ahaexecutivedirector@gmail.com
Website: www.americanhippotherapyassociation.org

American Hippotherapy Association Inc. (AHA, Inc.) promotes the use of the movement of the horse as a treatment strategy in physical, occupational and speech-language therapy sessions for people living with disabilities. Hippotherapy has been shown to improve muscle tone, balance, posture, coordination, motor development as well as emotional well-being.

AHA consists of medical professionals (physical, occupational and speech-language therapists) and others who are interested in the use of equine movement as a treatment strategy. Its philosophy includes the following:

- The organized movement of the horse is an effective treatment strategy used to improve function in patients with neuromusculoskeletal dysfunction.
- A licensed physical therapist, occupational therapist or speech-language pathologist with specialized training in the principles of hippotherapy and horsemanship is responsible for directing the treatment team during a session using hippotherapy as a treatment strategy.
- A hippotherapy session requires a carefully trained team consisting of: patient, therapist, horse expert, horse and sidewalkers.
- Equine Assisted Therapy, used by a medical professional, is part of a comprehensive treatment plan that integrates gains into functional activities.
- Horses used in Equine Assisted Therapy should have quality of movement, appropriate temperament and ongoing specialized training. All horses should be treated humanely at all times.
- Clinicians providing Equine Assisted Therapy are to maintain a high level of professional competence by participation in related continuing education activities. They must follow the standards of practice and code of ethics for their discipline.
- More research should be done to continue to validate the efficacy and effectiveness of hippotherapy and other Equine Assisted Therapies.

## AMET – Mexican Association of Therapeutic Riding

Balcones de la Colina 25
Fracc. Balcones
Queretaro 76140
MEXICO
Contact: Blanca Mayol
Phone: 442 214 3906
Email: amet.mexico@gmail.com
equinoterapia.amet@hotmail.com
Website: www.equinoterapia.net

AMET has been serving children with disabilities for over 36 years and represents Mexico's equine therapy centers by integrating, training, certifying, and promoting them internationally.

AMET's mission is to provide hippotherapy for children, adolescents and adults with physical and mental disabilities, behavioral issues and / or learning disabilities as well as victims of accidents. The association seeks to contribute to a better understanding of the needs and intrinsic value of persons with disabilities and to promote better relations among people with disabilities and the rest of society.
Furthermore, AMET trains and assists other institutions interested in creating new Equine Centers and it coordinates, certifies and regulates the establishment and operation of those centers. AMET also represent and supports them, both to increase their international knowledge and to receive financial support.

AMET has given rise to more than 150 new centers in Mexico and assisted 14 municipalities in their state. Presently, they offer 94 Equinotherapy courses, covering topics ranging from therapy for small children, music therapy, finger painting to substance abuse, familial violence, Rett syndrome, autism and neurodevelopmental and other conditions that can be treated with equine therapy. Additionally, they have 11 courses in neurohabilitation and hippotherapy and just added to their curriculum Natural Horsemanship.

## ANDE-BRASIL – Associação Nacional de Equoterapia

Granja do Torto
70636-000 Brasília DF
BRAZIL
Contact: Vera Horne
Phone: (55) (61) 34687092
Fax: (55) (61)3468-8486
Email: verahorne@gmail.com
Website: www.equoterapia.org.br

The National Association of Equotherapy, ANDE-BRASIL, was founded in May, 1989, by former Cavalry officers of the Brazilian Army that had the acumen to observe and appreciate the gains that some children with special needs began to show after some classic equitation lessons in a group with ablebodied students.

Since, many concepts and structures have been improved not only at ANDE-BRASIL but also in regard to the equotherapy method created. They have been disseminated throughout the country and have reached about 300 equotherapy centers that maintain a formal link to ANDE-BRASIL.

ANDE-BRASIL has invested in training health and education professionals with a 'Basic Course' (42hs) and an 'Advanced Course' (81hs) in equotherapy and by offering an equotherapy curriculum to riders with different degrees of technique in the field. We are looking forward to having teams formed at each of the equotherapy centers with the help of psychologists, physical therapists and equitation professionals.

ANDE-BRASIL has also organized congresses and conferences that are intended to bring together professionals from Brazil and from other countries every two or three years, including the International Congress of the FRDI (now HETI) in Brasília, in 2006. ANDE-BRASIL is now planning the next congress which will be the VII Brazilian Congress of Equotherapy, in 2017.

## ANIRE – Italian National Association of Therapeutic Riding

Via Bartolomeo D'Alviano 7
Milano 20146
ITALY
Contact: Daniele Nicolas Citterio
Phone: 39 03 3927 63674
Email: info@cnranire.it
Website: www.cnranire.it

ANIRE is the national association which introduced riding rehabilitation to Italy in 1976 and it is still the only recognized association of its kind. It deals with rehabilitation or riding therapy, and regards riding from a purely rehabilitative angle as in helping people achieve greater self-sufficiency. However, this does not rule out taking part in competitive sports activities. ANIRE was founded with the aim of rehabilitating people with physical and mental challenges with the help of horses and of promoting riding for recreational and sports purposes.

ANIRE also promotes scientific research, organizes and promotes riding rehabilitation in Italy, and trains specialists in riding therapy.

Since its foundation, the association's membership has grown to 10,000, encompassing 5,000 users/customers, 1,980 specialist sector workers (650 of which are currently active) and 170 centres spread over 20 regions, serving 12,000 patients with over 350 horses. Increasingly, centres run by municipalities, local health centres, regional governments and traditional rehabilitation centres, such as hospitals, enter agreements with ANIRE.

ANIRE was conferred the Milan Provincial Government Gold Medal and received a Certificate of Merit from its local municipality.

## ATE – Association Therapie Equ

13, Rue Mausereck
L-3934 Mondercange
LUXEMBOURG
Contact: Christiane Betz
Phone: 00352-55 26 15
Fax: 00352-26 55 00 65
Email: secretariat@atelux.lu
Website: www.atelux.lu

A.T.E.
ASSOCIATION
THERAPIE
EQUESTRE

Created in March 1982, ATE has been renting a small equestrian centre since 2004 where eight horses, three little ponies and two donkeys are housed. The association includes three physiotherapists with a certificate in hippotherapy and fifteen aids who lead the horses and help secure the patients, sometimes working with 3-4 horses simultaneously. Presently, the centre serves 106 clients.

Clients are children and adults with physical and mental challenges. In the educational and social part of their program, instructors work with clients with attention deficit disorder, sensory integration disorders, or behaviour issues.

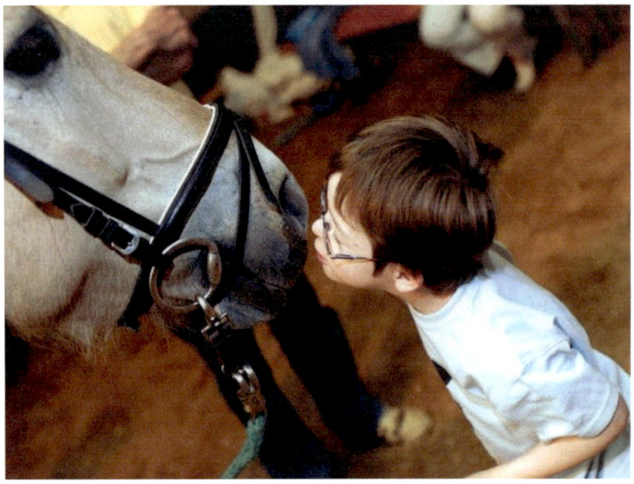

ATE works with clients individually or in groups of two, three or five, offering psychomotor exercises in form of vaulting or plays with psychomotor material in time and space. Clients learn to ride in groups with four horses until they are able to be integrated in a riding club outside.

## CanTRA – Canadian Therapeutic Riding Association

5420 Hwy. 6 North
RR#5 Guelph, Ontario
CANADA, N1H 6J2
Contact: Brenda Coward, Administrator
Phone: 519-767-0700 Fax: 519-767-0435
Email: ctra@golden.net Website: www.cantra.ca

CanTRA's mission is "to promote the benefits of therapeutic riding throughout Canada, by raising public awareness, working closely with the medical profession, providing continuing education, maintaining high standards for Therapeutic Riding Instructor Certification and accrediting therapeutic riding centres."

CanTRA is the national governing body for therapeutic riding in Canada, recognized by Equine Canada and HETI. Over 80 accredited member centres operate across Canada, working with children and adults who encompass a wide range of special needs.

CanTRA was founded in 1980 and delivers programs which include

- an internationally recognized Instructor Certification system with a high quality of service delivery
- an Accreditation system for therapeutic riding centres to ensure safe and high operating standards
- Hippotherapy training workshops for therapists
- an Equine-Facilitated Wellness program for certification of mental health, education and equestrian professionals working with mentally and emotionally vulnerable populations
- opportunities for Education through newsletters, clinics, workshops and national conferences
- a national Insurance Policy available to member centres.

## Centro di Riabilitazione Equestre Vittorio di Capua

A.O. Ospedale Niguarda Ca' Granda
Piazza Ospedale Maggiore 3
20162 Milano ITALY
Contact: Annalisa Roscio
Phone: 39 02.6444.2644
Email: riabilitazionequestre@ospedaleniguarda.it
Website: www.riabilitazionequestre.it

Vittorio di Capua was established in 1981 as part of the Department of Neuropsychiatry for Children and Adolescents at the Ospedale Niguarda Ca' Granda in Milan. It is the only organization of its kind that is officially recognized and approved by the national health authorities. The Center has been arranged to look like a farm, quite unusual in a large urban hospital, and is supported by equine staff trained to provide high level performance and guarantee low stress levels. The horses are also monitored by a therapeutic riding expert trainer and by veterinary staff.

The Center is currently treating 150 children with disabilities ranging from cerebral palsy to mental retardation, autism, and other syndromes. Staff includes one neuro-psychiatrist, one physiatrist, one psychologist, four therapists with different specializations, four assistants in therapeutic riding and over thirty volunteers. Treatments offered are free of charge. For each patient the medical rehabilitation team develops a personalized therapeutic program with medium and long term objectives and linked to the patient's condition, disability and correlated needs. Each session lasts around 30 minutes; afterwards the patient moves on to occupational-therapy sessions.

The Center's principal activities are:

- hippotherapy
- therapeutic riding
- pedagogic riding
- therapeutic vaulting

The Center is also involved in didactic and scientific activities such as specialized courses for therapeutic riding therapists and assistants and it collaborates with universities by offering internships for students and by providing staff members as speakers to conferences and other events.

## CHS – Czech Therapeutic Riding Association

Zemedelska 1
Brno
613 00
Czech Republic
Contact: Vera Lantelme
Phone: 00420 723 698 750
Email: info@hiporehabilitace-cr.com
Website: www.hiporehabilitace-cr.com
Facebook: www.facebook.com/Hiporehabilitace

ČESKÁ HIPOREHABILITAČNÍ SPOLEČNOST
1991 25 LET HIPOREHABILITACE JAK MÁ BÝT 2016

The Czech Therapeutic Riding Association (CHS) was founded in 1991 as a voluntary, non-government and non-for-profit organization with the vision to promote the best and safe practices of Equine Assisted Activities and Therapies (EAAT) and to integrate individuals with disabilities or special needs.

The Czech Therapeutic Riding Association registers 36 EAAT centres and 70 individual members.

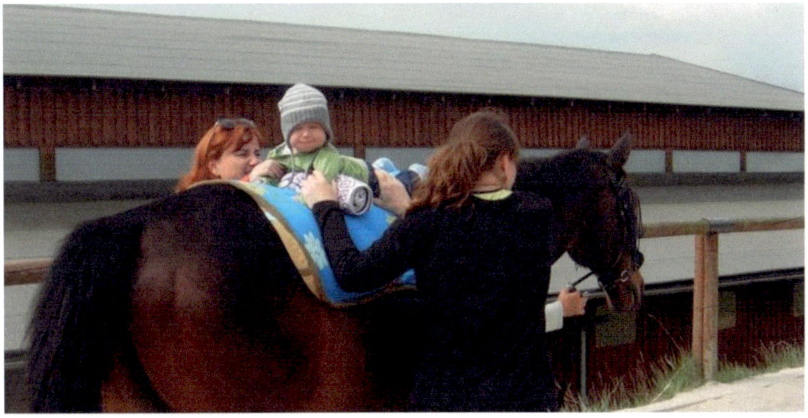

To promote best practices, CHS has implemented a Centre accreditation program and a Therapy horse certification program. The association organizes educational courses in Czech and English.

## CNRAFA ONLUS – Italian National Centre of Research, Therapy and Training

Via Bartolomeo D'Alviano 7
Milano 20146
ITALY
Contact: Daniele Nicolas Citterio
Phone: 39 03 3927 63674
Email: info@cnranire.it
Website: www.cnranire.it

Acknowledged by decree of the President of the Italian Republic as the national Italian association of horse therapy and of sporting and recreational riding for people with disabilities, CNRAFA is certified according to the international parameters. A provider of the Ministry of Health for continuous medical education, the organization has an agreement with the Ministry of Education, Research and University and holds a convention with Mondino, a mental health clinic for the evolutional age belonging to the University of Pavia.

CNRAFA includes:

POLYVALENT CENTRE:
Horse therapy; Psychiatric area; Equestrian rehabilitation; Cognitive area ; Re-instatement; Working re-instatement; Sporting re-instatement. 148 patients, 298 treatments a week, 1.138 treatments a month, 11.380 treatments a year.

POLYVALENT TEAM:
Consultant in: Neuropsychiatrics; Physiatrics; Sports Medicine; Psychologist ; Physiotherapist; Motory sciences graduate; Motory sciences and techniques of the motory preventive activities graduate; Teacher; Instructor of equestrian sports; Nurse; Vet.

THE STRUCTURE:
Manège; Office; Waiting room; Dressing room; Saddlery; Stable; Human and veterinary infirmary; Record office; Didactics room; Farrier workshops; Stable and grooming workshops; Maintenance workshops; Film library workshops; Computing workshops; Gardening workshop.

## CTEA – Chinese Taipei Equestrian Association

20, Chu-Lun Street, Room 808
10489 Taipei, Taiwan
Contact: Uta Rindfleisch-Wu
Phone: +8862/8771-1508
Fax: +8862/2778-3740
E-mail: ctea@ctea.org.tw
Website:www.ctea.org.tw

CTEA was founded in 1973. The main tasks of the association are to enhance the quality of horsemanship in Taiwan, to train coaches of different levels, to organize equestrian competitions and to choose riders for international competitions. CTEA has around 400 personal members and 35 group members.
Taiwan is an island where horses were not indigenous and horse sports therefore do not have any traditions. English style riding is the common riding style in Taiwan. While jumping competitions still draw the most participants, dressage is more and more on the rise.

In 2006, when Para Equestrian became a member of the FEI family, this part of therapeutic riding was introduced to CTEA and in March 2008 the first PE competition was held as part of a CTEA Dressage Competition. Since 2009 CTEA got a director of PE-Sports. In 2014, Sun Yu-Jen, a Grade II rider, was the only athlete from Taiwan joining the World Equestrian Games in Caen. He has also achieved Paralympic qualifying scores, which unfortunately did not translate into a ticket to RIO.

CTEA was proud of hosting the 2015 HETI Congress in Taiwan – and the very first in Asia – when representatives from 30 countries gathered on the campus of Aletheia University in Danshui. As a result of the Congress several Task Forces were set up, which will present the result of their work at the 2018 HETI Congress in Ireland.

## Danish Ridefysioterapi

Vesterkaervej 12
7600 Struer
DENMARK
Contact: Hanne Dahl Iversen
Phone: +45 22942270
Email: hdi@ridefysioterapeut.dk
Website: www.ridefysioterapeut.dk

DANSKE FYSIO TERAP EUTER

FRAKTIONEN AF RIDEFYSIOTERAPEUTER

The Center for Riding Therapy and Riding for the Disabled is a subsidiary of the Association of Physiotherapists in Denmark. An independent equestrian center, it is approved to practice therapeutic riding for disabled children and adults. Its facilities include:

- Riding stables
- Outdoor riding arena
- Scenic surroundings
- Heated rider room
- Accessible toilet
- Disabled parking and access,
- Climbing Ramp and lift and other equipment
- Eleven well-trained horses and ponies

The Center is open to anyone eligible for physiotherapy which includes diagnoses of congenital or hereditary diseases, acquired neurological disorders, physical disability due to an accident, reduction of context - and / or muscle function as a result of inflammatory rheumatic diseases as well as the blind and people with Autism or Down's Syndrome.

Instruction is provided by a disability riding instructor and an assistant. Tuition is free up to a maximum of 40 hours.

## DKThR – Deutsches Kuratorium für Therapeutisches Reiten E.V.

Freiherr-von-Langen-Str. 8a
48231 Warendorf
GERMANY
Contact: Almut Schlingenkötter
Phone: 0049 (0) 2581/927919-0
Fax: 0049 (0) 2581/927919-9
Email: DKThR@fn-dokr.de
Website: www.dkthr.de

With nearly 3000 members and extensive connections with various professional associations, the German Association for Therapeutic Riding (DKThR) plays a leading role in therapeutic riding in Germany. As the professional association for all occupational groups and institutions, DKThR is responsible for the promotion and support of all professionals, hospitals, therapy centres, equestrian clubs and schools associated with therapeutic riding.

An important task is the development of a scientific basis and promotion of scientific publications as is the cooperation with government and private purchasers. Its international role is reflected in cooperation with FRDI and FATP. DKThR freely shares its know-how with other countries and furthers the exchange of information.

During the past 40 years DKThR has also established itself in the field of education and training of professionals and, in cooperation with the NF, developed quality management and quality standards. DKThR offers professional education in four fields of Therapeutic Riding:

- Hippotherapy
- Psycho-educated vaulting and riding
- Riding as sport for people with disabilities
- Occupational Therapy with the horse

Finally, DKThR supports the high-performance sport for disabled in Para-Equestrian –Dressage-Driving and is a member of the Special Olympics.

## ELETHIP – Hellenic Association for Therapeutic Riding

18 Karaiskaki Str.
TK 19400 Koropi, Athens
GREECE
Contact: Irene Kapari
Phone: 0030 6977 20 11 86
Email: irenekapari@hotmail.com
Website : www.elethip.gr
Facebook: //web.facebook.com/ΕΛΛΗΝΙΚΗ-ΕΤΑΙΡΙΑ-ΘΕΡΑΠΕΥΤΙΚΗΣ-ΙΠΠΑΣΙΑΣ-ΕΛΕΘΙΠ-141088019295215/?ref=bookmarks

ELETHIP is a non-profit association that started its journey in the field of Equine Assisted Activities and Therapies in 2003 in Athens – with one horse, one PT and one rider. Since, ELETHIP grew to provide its services to 90 riders a week with a therapeutic team that consists of 8 certified professionals in various fields (such as physical therapists, TR & dressage instructors, occupational & speech therapists, psychologist and social worker) and 8 horses. The team also involves 15 trainees per year and 18 well trained volunteers.

The mission of ELETHIP is:

- to provide equine assisted activities and therapies with safety to people with disabilities from the age of 2
- to provide the experience of riding as a sport with our athletes competing at national and international Special Olympics & para equestrian dressage competitions
- to provide certified professional education in all fields of therapeutic riding through workshops, seminars and an annual conference
- to conduct & publish evidence based research about the benefits of equine therapy
- to increase public awareness for people with disabilities and the welfare of the horse
- to cooperate productively in a regular basis with other national and international organizations that are actively involved in the same areas (equines, TR, hippotherapy, physical & mental health, sports, etc.).

# FCA – Fundacion Caballo Amigo

Lope de Vega, Nº 36, 5º D
28014 Madrid
SPAIN
Contact: Mercedes Jiménez Horwitz
Phone: 34 910 228 219
Email: gestion@caballoamigo.org
Website: www.caballoamigo.org
Facebook: www.facebook.com/fundacioncaballoamigo

In 1998, a group of professionals in medicine, education and sports founded the F.C.A. with the purpose of developing equestrian activities for disabled people in Spain. The foundation's activities focus on four main areas:

**Therapy**

Includes therapy, education, enjoyment, and social integration for people with disabilities with horses and riding. FCA is presently working with disabled riders with a variety of different disabilities.

**Education**

Courses in therapeutic riding are designed to train professionals from medicine, education and sports. FCA also offers a two year program in conjunction with the Complutense University of Madrid and the University of Bovigny (France).

**Expansion**

Includes various activities of the foundation with major emphasis on the scientific investigation of therapeutic riding.

**Sport**

FCA is responsible for the development of para-equestrian activities at the Spanish Equestrian Federation.

## FCRT – Fortune Centre of Riding Therapy

Avon Tyrrell
Bransgore
Christchurch, New Forest, BH23 8EE
UNITED KINGDOM
Contact: Mrs. Jennifer Dixon-Clegg
Phone: 44 1425 673297
Fax: 44 1425 674320
Email: enquiries@fcrt.ac.uk
Website: www.fortunecentre.org

The Fortune Centre of Riding Therapy (Registered Charity Number: 1045352) is a place that provides education and therapy through horses for people that are more motivated by being with horses than anything else. Weekly sessions are run for local children with a variety of disabilities. A day care provision exists for local people with special needs who wish to focus on working with horses. FCRT provides work experience placements for allied professionals usually during their professional training in physiotherapy, occupational therapy, teaching or nursing.

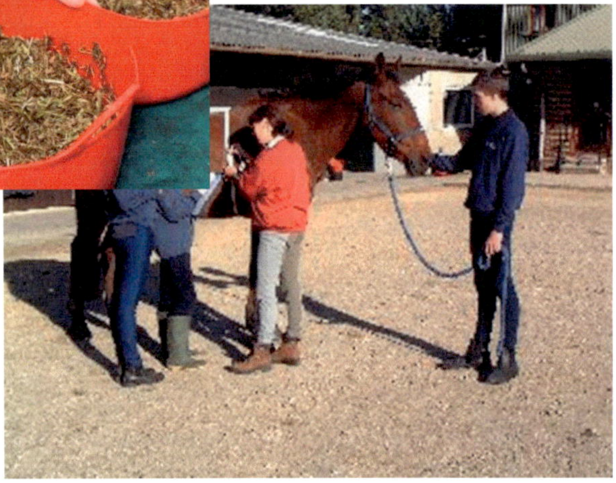

The FCRT's main activity is as an Independent Specialist Provider for young people, aged 16-25, with learning difficulties and disabilities whose educational needs cannot be met in government provision. The residential course *Further Education Through Horsemastership* (FETH) runs over 3 years for 45 students at any one time from anywhere in the UK and occasionally from overseas.

## Fédération Nationale Handi cheval

26 Route du Pre Madame
Royeres 87400
FRANCE
Contact: Glen Gouttes
Phone: 33 555 69 21 98
Email: fnhc@wanadoo.fr
Website: www.handicheval.asso.fr

## FENTAC – Fédération Nationale de Thérapie avec le Cheval

12 Allee du Parc
89170 St. Fargeau
FRANCE
Contact: Dominique Darques
Phone: 33 86-74-0378
Email: darques.dominique@orange.fr
Website: www.fentac.org

The National Federation of Therapy with the Horse is a non-profit organization created in 1986. FENTAC aims to:

- bring together groups and individual practitioners of therapeutic riding
- develop and oversee enforcement of specific guidelines for Therapy with the Horse
- promote research regarding the theoretical and practical aspects of this therapy
- support other organizations associated with Therapy with the Horse
- take part in meetings and conferences in France and abroad which relate to therapeutic riding

## Festina Lente

Old Connaught Avenue
Bray
Co Dublin
IRELAND
Contact: Jill Carey
Tel: 000353012720704
Email: info@festinale Website: www.festinalente.ie

Festina Lente is a registered charity which provides a wide range of equine based programmes to young people, teenagers and adults with physical, intellectual, psychosocial, behavioural and/or mental health difficulties. These programmes include:

- Three Year Accredited Vocational Equestrian Training for young people with disabilities aimed at accessing employment in the equestrian industry or further training
- Equestrian Employment Programme for adults with disabilities
- Supported Helper Programme for young children and teenagers helping in the Equestrian Centre with additional supports, for example, staffing ratio 1:1 or 1:2
- Equine Assisted Learning to support emotional, cognitive and/or behavioural changes amongst young people and adults
- Equine Facilitated Educational Programme for young children affected by socio economic disadvantage
- Therapeutic Riding for the development of physical, psychological and /or psychosocial well-being of young people and adults with disabilities

Festina Lente provides two Quality in Qualifications Ireland (QQI) nationally accredited programmes, a Therapeutic Coaching Training Programme and an Equine Facilitated Assisted Learning Programme, and engages in a variety of research projects focusing on the benefits of horse human relationships, horse welfare and management considerations. A core value of Festina Lente is the need to achieve a balance between people's needs, expectations and choices of equine based programmes and the equines needs, a big challenge in terms of economics and human resource management. Festina Lente strives to implement the International Society of Equitation Science (ISES) principles in all its interactions with equines. All the equines involved in Therapeutic Riding wear Micklem Multipurpose Bitless Bridles or Head collars.

## FISE – Federazione Italiana Sport Equestri / Italian Equestrian Federation

Viale Tiziano 74
Rome 00196
ITALY
Contact: Barbara Ardu
Tel: 39 06 83668427
Email: riabilitazione@fise.it
Website: www.fise.it

The Italian Equestrian Sports Federation (FISE) was founded in Rome in 1926 and, approved by the Italian Olympic Committee and the Ministry of Culture, the Federation is the only authorized organization to regulate equestrian activities in Italy whether they are of an educational, athletic, recreational or training nature.

FISE has a specific Department for Therapeutic Riding. Its goal is to develop opportunities for people with disabilities so they can attain treatment through Therapeutic Riding.

Every year, FISE organizes special training courses for riding instructors and therapists in various areas of social health. The Therapeutic Riding Department also conducts special research in the mental health field, especially on the use of therapeutic riding in psychiatric rehabilitation.

Finally, the Therapeutic Riding Department also organizes congresses, seminars, and workshops to improve and promote Therapeutic Riding.

## FN – Fédération Equestre Nationale / Deutsche Reiterliche Vereinigung e.V

Freiherr-von-Langen-Straße 13
48231 Warendorf
GERMANY
Contact: Soenke Lauterbach
Phone: 49 (0) 2581/6362-0
Fax: 49 (0) 2581/62144
Email: fn@fn-dokr.de
Website: www.pferd-aktuell.de
Facebook: www.facebook.com/DeutscheReiterlicheVereinigung
Twitter: www.twitter.com/FN_eV
Youtube: www.youtube.com/FNPferdesport

Established in 1905, the German Equestrian Federation (FN) is one of the largest national equestrian federations in the world. Within the German Olympic Sports Association (DOSB), it ranks as Germany's 9th largest sports association. The FN's duties and responsibilities are distributed to three separate pillars: Sports, Breeding and Personal Members. In the sports sector, in collaboration with the respective member associations EWU (Western Riding), DKThR (Therapeutic Riding and Para-Equestrian) and VDD (Endurance), the FN takes care of the disciplines Dressage, Jumping, Eventing, Driving, Endurance, Vaulting, Reining and Para-Equestrian.

The FN is the umbrella organisation for approximately 7,600 riding clubs and 4,000 commercial equestrian centres with a total of 709,000 members, all being represented by 17 regional equestrian associations. In addition, 25 breeding organisations are member associations of the FN with about 3,500 riding horse sires and 55,600 riding horse brood mares.

FN regularly issues uniform guidelines, rules and regulations, one of them being the APO. Among other items the APO contains details on the training programme for specialised personnel in Therapeutic Riding. These are the basis of the educational programme of the DKThR (German Curatorship for Therapeutic Riding). DKThR cooperates with FN in education, labelling and qualification of riding and therapy facilities for the Disabled.

## FPG – Federatie Paardrijden Gehandicapten

Wekeromseweg 12
6816 VS Arnhem
THE NETHERLANDS
Contact: Dieuwke Papma
Phone: 0031-6-46745415 (Tuesdays and Thursdays)
Email: info@verenigingfpg.nl
Website: www.verenigingfpg.nl

Since 1968, FPG is the Dutch organization of riding for people with disabilities. It is a certified fundraising organization. FPG has seventy-five members, all local groups which offer horse riding lessons on a non-profit basis, but working on a professional level. For a list of members, please look at www.verenigingfpg.nl/ledenmaneges.php.

The Objectives of FPG are to provide quality and safe riding opportunities for persons with disabilities, for recreational, sportive as well as therapeutic reasons. Activities include:

- education of riding instructors (qualification level 3) by the National Equestrian Sports Federation KNHS
- education in vaulting for disabled
- refresher courses for instructors
- courses for volunteers
- programs on safety- and risk management in FPG riding schools – all members are certified as safe riding schools.
- development of quality and safety standards for EAT
- integration of activities with the National Equine Sports Organization, KNHS, and the Federation of Riding Schools, FNRS.

FPG wants to support the foundation of a national organization of professional equine assisted therapists, stimulate more scientific investigations into the effects of EAT and support development of an EAT-course for therapists.

## GUL – God Unlimited Outdoor Therapy

CIO number 1159337
The Bustard, Shrewton,
Salisbury, Wiltshire, SP3 4DU
UNITED KINGDOM
Contact: Dr C C Grummitt
Phone: 01980 621712
Email: office@god-unlimited.org
Website: www.god-unlimited.org

God Unlimited (Gul) was founded in 2006 to provide people from all walks of life the chance to enjoy the outdoors in a therapeutic setting. In our first year, with two Shetland ponies, at the home of our founder, Gul provided 200 hours of activities, in 2014 we provided over 10,000.

Gul now uses 4 sites based in Wiltshire, Hampshire and Wales. We offer Outdoor and Equine Assisted Therapies (EAT), including riding, exercise, pain management back to work programmes and education using staff, volunteers and our 14 ponies.

This incredible growth is due in no small part, to the skilled team of clinical staff and dedicated volunteers. Gul is able to offer therapy to all sorts of people, both children and adults, able bodied and not and also provide for military families and our local community. Gul aims to increase the independence and quality of life for everyone who uses our sites and accepts referrals from local authorities, education providers and individuals.

## HANDHIPPO – Belgische Nationale Vereniging van Therapeuten Gespecialiseerd in Hippotherapie / Union Nationale Belge des Thérapeutes Spécialisés en Hippothérapie

Rue de Beriot, 33/Maatschappelijke zetel: de Beriotstraat, 33
1210 Bruxelles-Brussel
BELGIUM
Contact: Marie-Thérèse Kuypers
Phone: 32/ 498 48 32 44
Email : handhippo@skynet.be
Website: http://users.skynet.be/handhippo

Founded in 1974, the national Belgian non-profit organization Handhippo is composed of therapists specialising in therapeutic riding. Their definition of therapy as a medical treatment in the broader sense covers all aspects of the patient, both physical and mental, and the horse represents everything that it brings by his presence, his relationship, his contact which is both rewarding and special.

Therapy is given under prescription or on medical indications.

Handhippo's goal is to improve the credibility of everyone's work and to present a coherent image of the activities associated with therapeutic riding to the different professions, the media and the public.

## Hest og Helse

Gml Enebakkvel 71
1188 Oslo
NORWAY
Contact: Annika Meltzer
Phone: +47 928 34 842
Email : annika.meltzer@getmail.no
Website: www.hestoghelse.no

Hest og Helse, Horse and Health, is a voluntary Norwegian national organization which was established in 1971. Currently they have about 300 members – individuals and businesses – and a number of well-qualified course instructors with extensive experience.

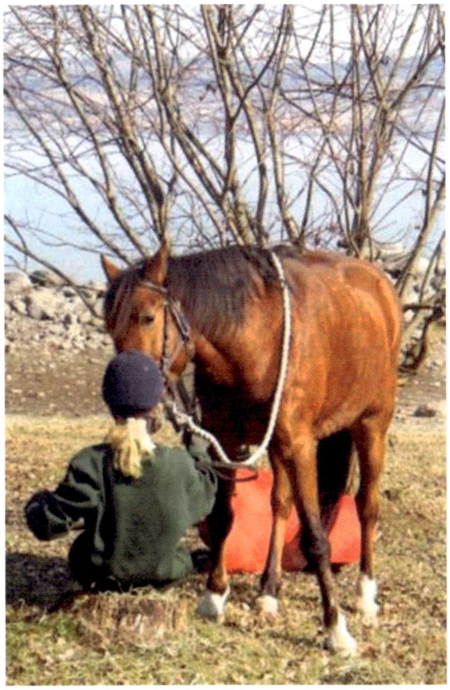

Hest og Helse's extensive training portfolio deals with the horse in health care and the exciting intersection of horses and human health. It includes:

- horses in psychiatry and social medicine
- disability management
- training in horse therapy
- management of children and adolescents

The organization's purpose is to promote the use of horses in the health context. Hest og Helse strives to actively contribute to the optimal quality in the use of horses to improve the physical, mental and social health of its clients.

## HETRICS/KETHIS – Hellenic Therapeutic Riding Center of Serres

N. Skopos
Serres, Macedonia 62100
GREECE
Contact: Maria Liga
Phone: +30 6945549696
Email : kethiser@gmail.com
Website: www.kethis.gr

HETRICS, the Hellenic Therapeutic Riding Center of Serres, is a non-profit organization founded in December 2005. The main goal of the center is to promote the idea of riding as a means of therapy and to advance the inclusion of people with disabilities into the social community. Anyone who wishes to be engaged with riding either as therapy or as a sport or simply to enjoy a ride in the country side will love our center.

Children and adults can take riding lessons with the assistance of a qualified horse trainer. Another activity of the center is organizing therapeutic riding seminars. Our therapeutic team consists of two physiotherapists, two teachers of gymnastics, a psychologist, a special pedagogue, a social worker, a speech therapist and an occupational therapist. We all are volunteers!!

The center is located in Northern Greece, in Serres, a town which is 80 km outside of Thessaloniki, the second capital of Greece. HETRICS has taken part in the World Special Olympics Games twice - Shanghai 2007 (4 bronze medals) and Athens 2011 (2 gold, 3 silver and 5 bronze medals).

## Horse Community Piskari Japan

327-9 Nishi-horobetsu, Urakawa-cho
Urakawa-gun, Hokkaido
057-0002
JAPAN
Contact: Katsuhiko Sumii
Phone: +81-90-3629-2842
Fax: +81-146-28-1305
Email: piskari.joba@gmail.com
Facebook: www.facebook.com/piskari.joba/

We are based out of Urakawa, a town located in Japan's northernmost island of Hokkaido. Urakawa is located in the Hidaka area of Hokkaido. Its prosperous equine industry and ideal climate for breeding and raising horses, provides Japan with 70% of its racehorses. The Horse Community's riders range in age from 2 years to 95 years old. The Horse Community is working in conjunction with the welfare department of the Urakawa government center. Urakawa's local government is the only entity that is currently supporting horse therapy in all of Japan.

Therapeutic riding at the Horse Community Piskari Japan is called "riding conductive education". It provides people with mental and physical disabilities an opportunity to touch and be with horses and to ride horses, as well as a chance to maintain and improve their abilities while enjoying themselves. Therapeutic riding is also used to promote participation in society through horseback riding. Therapeutic riding instructors and physical therapists are involved with the riding lesson at the Horse Community.We are also researching the effects of Therapeutic Riding:

- Changes in pressure distribution on seat before and after riding therapy
- Considering the mental effects of riding therapy
- Effects of riding therapy - example of application to the physical therapy of a subject with partial paralysis on the right side.

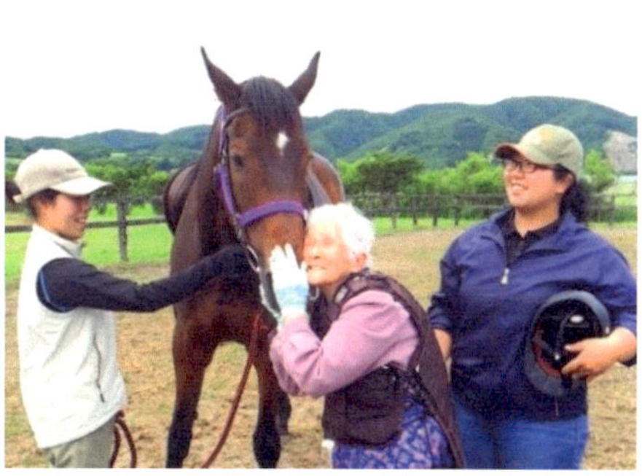

## Hungarian Riding for the Disabled Federation

MAGYAR LOVASTERÁPIA SZÖVETSÉG

24. Terez Krt.
Budapest 1066
HUNGARY
Contact: Peter Edvi, President, Gabriella Bozori, First Vice President
Tel : 361 475 7000
Email : edvi@GYERMEKMENTO.HU
Website: www.lovasterapia.hu

The Hungarian Riding for the Disabled Federation was founded in 1997 with the patronage and support of the International Children's' Safety Service. At the time, no nationally acknowledged and accredited organization existed to provide courses in therapeutic riding. The first training was held by Sister Chiara Hatton Hall and seven English specialists. The following years Prof. Dr. Med. Carl Klüwer, Octavia Brown, Anita Shkedi, Dr. Danielle Citterio, Peter Holzmüller, Susanne von Dietz, Isabelle V. Neumann-Cosel, Kurt Lehmann, Willem van Lieshout, Reinhard Tötchinger, Gundula Hauser and Sharon Herz-Freilich held training sessions.

The Hungarian Ministry of Education and the NFFH (The National Bureau of Andragogy and Vocational Training) accredited or acknowledged the Hungarian Riding for the Disabled Federation's training. The Federation provides 2 year training for four different specializations:

- Teachers of children with special needs – (leader: Gabriella Bozori)
- Physiotherapists (leader: Peter Holzmüller)
- Psychologists (leader: Willem van Lieshout)
- Para-sport trainers (leader: (interim) Gabriella Bozori)

The Federation is also involved in helping the Hungarian Parliament and ministries prepare laws pertaining to therapeutic riding, such as riding services in Hungary and the Hungarian sport law decree which affects para-sport. Since 2005, the Federation organizes a Competition for Disabled Riders each year, together with the International Competition (CDI***). These events have become very popular and help promote the integration of people with special needs. The Competition is held at the end of September every year, and we also organize a Conference for professionals around 23 October. Groups, presenters and guests are welcome for both events.

## Ippostirixi – Therapeutic Riding, Hippotherapy and Education

Lagadas 57200
Thessaloniki
GREECE
Contact: Katerina Rokou
Phone: 0030 6948727166
Email: ippostirixi@yahoo.gr
Website: www.ippo-stirixi.blogspot.gr

Ippostirixi (Horse Support) was established in March of 2013 as a nonprofit company with a mission to practice and promote equine assisted therapy. Its purpose includes the treatment and education of people with disabilities through equines and other animals as well as the creation and implementation of educational seminars for the training and certification of stakeholders (educators, psychologists, physiotherapists, physical education teachers, speech therapists, occupational therapists, mental health counselors, etc.) in educational and therapeutic horseback riding and hippotherapy.

At Ippostirixi right now we run a therapeutic and educational program for children and adults that face emotional, behavioral, mental, neurological and motor difficulties. The lessons may be individual or in a group, depending on the participants' needs and goals. Instructors, trainees and volunteers run the program.

Ippostirixi has no employees, all our staff are volunteers. Our instructors have years of training as therapeutic riding instructors and are psychologists and physiotherapists. Trainees and volunteers aid during the lesson as leaders or side walkers. Ippostirixi owns 3 horses that have gone through a solid training as therapeutic riding horses.

## JTRA – Japan Therapeutic Riding Association

c/o Iwai Medical Foundation
3-17-6 Higashi-Shinagawa-ku
Tokyo 140-0002
JAPAN
Contact: Seiji Miyasaka
Phone: 81-3-3565-6641
Email: office@jtranet.jp
Website: www. jtranet.jp

The Japan Therapeutic Riding Association was established as a Japanese non-profit organization in March 2007, followed by the certification by the Governor of Tokyo Metropolitan Government where the principal office of the Association is located.

The aims and purposes of the organization and membership include establishing and promoting sound development of therapeutic riding in Japan through activities such as the following:

- developing the practical methods of therapeutic riding
- evaluating the effects of therapeutic riding
- conducting practical seminars and public relations activities for therapeutic riding
- coordinating and collaborating with domestic and international associations concerning therapeutic riding
- breeding, raising and training of horses that are used for therapeutic riding
- training the staff for therapeutic riding
- introducing reference books, DVDs, tack and other materials used for therapeutic riding

## KATH – Korean Academy of Therapeutic Horsemanship (RDA - Korea)

364, Gunpo-ro, Gunpo-si, Gyeonggi-do
SOUTH KOREA
Contact: Amy Ji Young Lee
Phone: 82-31-461-6637
Email: kathkr2013@gmail.com
Website: www.kath.kr

The Korean Academy of Therapeutic Horsemanship promotes and advocates professionalism and scientific approach in equine-assisted activities and therapies by providing variety educational opportunities and events.

Main activities:

- Promoting the academic and research activities related to therapeutic horsemanship
- Development and propagation of professional training and education
- Establishing friendship between members and with governmental and nongovernmental organizations
- Improving public awareness

We are dedicated to improve health and to enhance the quality of life for mankind through the provision of the state-of-the-art patient care, clinical research and education.

# LAPO – Italian Association of Families, Bodies and Professions against Neurological and Psychiatric Diseases in Childhood and Adolescence

Via Antonio Susini n.14
50125 Firenze
ITALY
Contact: Prof. Anna Pasquinelli
Phone: 39- 347.5398994
Fax: 39-55.4298432
Email: anna.pasquinelli@unifi.it; associazione-lapo@gmail.it
Website: www.associazione-lapo.it

The Italian Association of Families, Bodies and Professions against Neurological and Psychiatric Diseases in Childhood and Adolescence (LAPO) was co-founded by Prof. Massimo Papini and Prof. Anna Pasquinelli. Its activities within the area of Therapeutic Riding focus specifically on the "Section for Study, Updating, Research in TR and Sport Riding for the Disabled". LAPO is responsible for organisation, information, professional advice, and supervision of therapeutic riding centres in Italy and maintains contacts with professional associations, public institutions, administrative and legislative bodies, and the National Equestrian Sport Federation. The association organizes congresses, conferences and workshops on therapeutic riding.

In collaboration with the Chair of Child Neuropsychiatry of the University of Florence LAPO developed the *University Curriculum in TR for Post-Graduate Physicians in Child Neuropsychiatry* (1995) and, in 1996, founded the Italian *Study Group on TR,* a multi-professional group with scientific goals which meets twice a year. LAPO also developed a *Theoretical/Practical Training Course in TR* in 1997 and a *University Master in TR* in 2002/2003.

LAPO promotes and contributes to the study, research and scientific experimentation on therapeutic riding and has, in collaboration with the University, developed a *TR Neuromotor Disorders Assessment Scale* for the quantitative assessment of the results on horseback and on the ground as well as a definition of the guidelines for therapeutic riding activities in a therapeutic riding centre. Published books include *Practical Principles of Therapeutic Riding* (1996) and a *Handbook of Therapeutic Riding: Principles, Methodology, Organisation* (2009; in press the English version in 2016). LAPO also provide a bibliography on therapeutic riding and technical aids for riding with disabled riders (Hippotherapy saddle "Elisa 2").

## Let's Run (formerly RDA-KRA)

Gyeongmagongwon-Daero 107
Gwacheon-si
Gyeonggi-do
427-711
SOUTH KOREA
Contact: Kim Soo Hyeon
Phone: 82 2 509 2149
Email: kra1064@kra.co.kr
Website: http://company.kra.co.kr/global/english/main

Let's Run (formerly RDA-KRA) joined HETI as a full member in August 2010 after three years of being an Associate member. The Korea Racing Authority (KRA) is a non-profit organization established by the KRA Law. It is the sole racing authority in Korea and is under the supervision of the Ministry for Food, Agriculture and Fisheries.

Let's Run's objectives are to authorize, operate horse racing and promote the equestrian industry in KOREA. One of the main objectives is to be a leading influence in charities and community services. Riding for the disabled is the main community service of Let's Run.

Let's Run started RD programming in 2005 (as KRA). At that time it was a pilot project because KRA had no access to a professional RD instructor or advisor here in Korea. Now, Let's Run has 2 RD instructors and one of them is also an Occupational Therapist as well. In 2010 they are providing hippotherapy and therapeutic riding programs with 5 therapeutic riding ponies for 108 clients.

In addition, Let's Run strives to be a leader in disseminating RD in Korea. Let's Run has hosted several PATH INTL. workshops & certification courses in the last 3 years. Many people are interested in this field and Let's Run wants to ensure that RD instructors have access to knowledge, practical experience and safety awareness.

## National Federation of Therapeutic Riding (Russia)

Weschnikowskoje House 1, 1, App155
Moscow 111402
RUSSIA
Contact: Alexander Ivanovich Denisenkov
Phone: 8-903-187-49-42
Email: rider960@yandex.ru

The National Federation of Therapeutic Riding is a community of people who've dedicated themselves to implement therapeutic horse-riding methods in Russia. To ensure TR receives the recognition it deserves and its rightful place among the rehabilitation areas, the Federation works to create conditions for the development and dissemination of equestrian sport for disabled individuals.

The National Federation of Therapeutic Riding and Equestrian Sport for People with Disabilities (NF of TR & ESPD) was founded on 8 November 1997 by the participants of the II International All-Russian Hippotherapeutic Conference and was officially registered on March 4, 1999.

The Federation organizes the preparation of TR instructors and specialists in Equestrian Sports for People with Disabilities (trainers, sport judges, physician classifiers, etc.). From 1994 to 2002 NF of TR & ESPD, together with the specialists of Georgian Medical Academy, prepared 122 specialists. The specialists received the international certificate of "TR and hippotherapy trainer in physical rehabilitation system".

Since 2003, the National Federation in conjunction with the Department of Preventive Medicine of FRQHW (Faculty of Raising Qualification of Health-care Workers) PFUR runs educational cycles "Urgent problems of complex rehabilitation of patients with disabilities based on TR&ESPD". More than 500 TR&ESPD specialists have been prepared over 12 years.

## Norwegian Physiotherapist Association

Norsk Fysioterapeut Forbund/Norwegian
Physiotherapist Association
Post Boks 2704 St. Hanshaugen
N- 0131 Oslo
NORWAY
Contact: Kari Liavag - Phone: +47 9520 1707
Contact: Mette Grav - Phone: +47 9510 0325
Email: grav@broadpark.no
Website: http://fysio.no/Forbundsforsiden/Organisasjon/Faggrupper/Ridefysioterapi/Om-faggruppen

Norwegian Physiotherapist Association, NPA, has more than 10.000 members, including the special interest group of riding physiotherapy with approximately 100 members. Scandinavian physiotherapists started using horses for therapy at the beginning of the 19th century. At present the Norwegian Physiotherapist Association offers a 100 hours continuous education program in therapeutic riding for physiotherapists who can document sufficient riding skills (beginners FEI level, Lett B Norwegian Equestrian Federation level) and horsemanship, e.g. The Norwegian Horse Centre's horsemanship courses for physiotherapist.

Riding physiotherapy is part of ordinary health care treatment in Norway, and patients need an approved referral from a physician/GP. Riding physiotherapy is a group-based treatment with 30 minutes of mounted activities in each session. Individual treatment, 30-60 minutes, is optional. Groups are made up by 2-8 riders, depending on needs. Walking assistants are used when necessary, e.g. horse handlers or next of kin. The National Insurance system reimburses patients for 30 therapeutic riding sessions a year, covering 30-50% of the cost. Other costs (stable fees, extra sessions) must be paid out of pocket.

The NPA riding physiotherapy interest group aim at encouraging research on therapeutic riding, to secure and enhance the quality of continuous education in therapeutic riding for physiotherapists, and to contribute to clinical and academic knowledge sharing. Acknowledging animal welfare is vital to further the practical and academic standing of riding physiotherapy.

## NZRDA – New Zealand Riding for the Disabled Association

Po Box 58110
Whitby
Porirua, 5245
NEW ZEALAND
Contact: Chris Hooper
Phone: 64 4 234 6090
Fax: 64 4 234 6094
Email: admin@rda.org.nz
Website: www.rda.org.nz

New Zealand Riding for the Disabled Association Inc.

Riding for the Disabled was first established in New Zealand 50 years ago. NZRDA is the umbrella organisation for its 56 member groups located throughout New Zealand. They provide over 3,000 children and adults with an opportunity for enjoyment, therapy, rehabilitation, freedom, friendship, laughter, and the achievement of independence and confidence with the help of the horse.

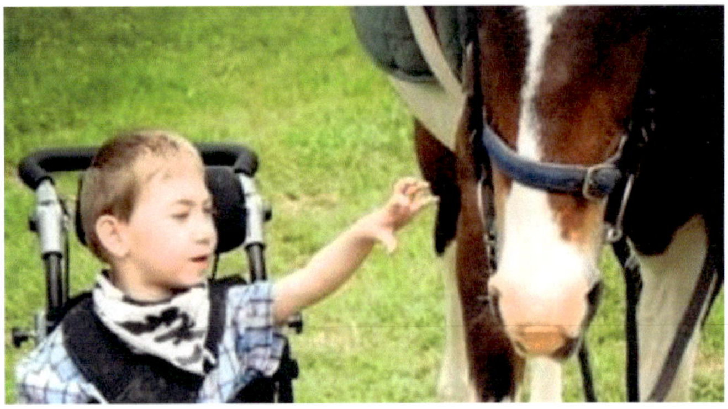

NZRDA offers advisory services, standardised training, and guides to the operational practices of its member groups. The RDA vision is: enhancing quality of life for those living with physical, intellectual, emotional and social changes. All RDA Groups are entities in their own right and offer a variety of services for their riders. RDA Groups offer individualised programmes according to the needs of their riders, but in general the following fields are covered by the RDA movement in New Zealand:

- Education- for riders with special needs, equine vocational training for high school students, as well as education and training for helpers, coaches and therapists.
- Therapeutic riding
- Rehabilitation
- Hippotherapy- NZRDA also offers qualifications for therapists and Horse handlers
- Interactive vaulting
- Psychotherapeutic riding
- Carriage driving, sport and recreation including support and assistance for Para-Equestrians, and Special Olympic riders.

# Österreichische Initiative – Pferde helfen Menschen in cooperation with e.motion

An der Niederhaid 4
A-1140 Vienna
AUSTRIA
Contact: Gundula Hauser
Phone: 43 699 11 555 884
Email: g.h.hauser@gmx.at
Website: www.pferd-emotion.at

The Austrian Initiative – Horses help Humans, in cooperation with the Association e.motion, offers advanced training for instructors and therapists in the field of:

- "The Healing Language of Horses". These courses are supported by the University of Vienna, the Sigmund Freud University. E.motion did a lot of scientific work in this field that was filmed by ARTE and published by international broadcasters.
- Equine assisted Therapy (or THR) by "Systemic Pedagogic"
- Counselling with the help of the horse

Tuition is charged for the courses and the courses are usually given in German, when necessary also in English. A certificate is always issued at the conclusion of the courses.

# OHI – Organisationen för Hästunderstödda Insatser

c/o Tillberg
Hjalmar Olsons v 16
134 40 Gustavsberg
SWEDEN
Contact: Pia Tillberg, chairperson
Phone: 46 70 403 71 09
Email: kontakt@ohi.nu
Website: www.ohi.nu

OHI, the Swedish Organisation for Equine Assisted Interventions (former IRT) was founded in 1994. It is a non-profit organization. OHI is a nationwide association and its full members are medical doctors, physiotherapists, occupational therapists, psychologists, nurses, psychotherapists, speech therapists, social workers and special education teachers. The organization is also open for anyone to join as an associated member.

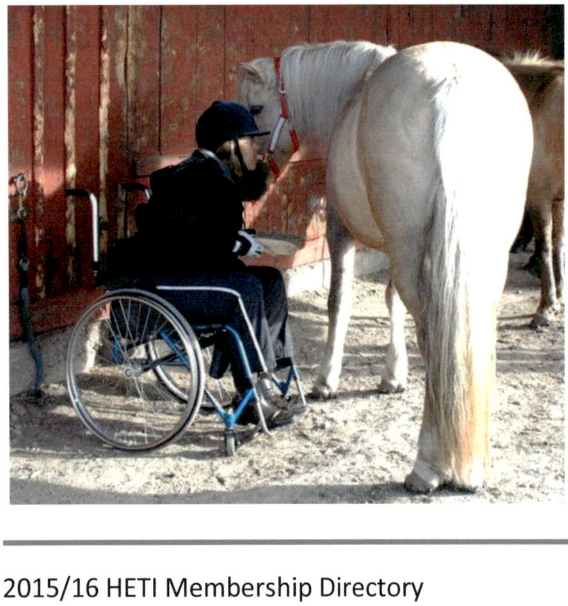

The OHI Newsletter is distributed four times a year. It contains information about courses, conferences and other material of interest.

OHI supports and conducts courses in equine assisted interventions. Since 2009 members have been able to apply for certification according to national standards and qualifications in regard to EAI and general knowledge of horses.

## PACTH – Pennsylvania Council on Therapeutic Horsemanship

346 Briar Lane
Chambersburg, PA 17201
USA
Contact: Ann O'Shallie
Phone: 717 552-1628
Email: aoshallie@comcast.net
Website: www.pacth.org

PACTH's mission is to provide mentoring, education, and certification in the field of Equine Assisted Activities and Therapy (EAAT).

The Pennsylvania Council on Therapeutic Horsemanship (PACTH) is a coordinating council serving therapeutic horsemanship programs and the participants, instructors, volunteers and administrators who are involved in these programs.

Founded in 1984 by Penn State University, Pennsylvania State 4-H and the Pennsylvania Easter Seal Society, the PACTH has a long history of support to the profession of therapeutic horsemanship through the following:

- Workshops & Seminars
- Pennsylvania Qualified Instructor Training Course (PQI)
- Annual Conference
- Networking opportunities
- Start-up Information
- Newsletter
- Exhibits at Public Functions
- State Host for PATH INTL. Region II Conference

## PATH International – Professional Association of Therapeutic Horsemanship International

P.O. Box 33150,
Denver, CO 80233
USA
Contact: Kathy Alm
Phone: 303 452-1212
Email: kalm@pathintl.org
Website: www.pathintl.org

PATH INTERNATIONAL
Professional Association of Therapeutic Horsemanship International

The Professional Association of Therapeutic Horsemanship International (PATH Intl.) was founded in 1969 as the North American Riding for the Handicapped Association to change and enrich lives by promoting excellence in equine assisted activities. Since its formation the association is no longer limited to North America; members provide many services in addition to therapeutic riding; and an infinite number of special needs, thankfully no longer thought of as handicaps, are known to benefit from the power of the horse.

With over 860 member centers and nearly 4,700 certified professionals around the globe, 61,832 children and adults find physical and mental well-being plus a sense of independence through involvement with equines. In addition to therapeutic riding, a center may offer any number of equine assisted activities including hippotherapy, equine facilitated mental health and psychotherapy, driving, interactive vaulting, trail riding, competition, ground work, and stable management.

Through a variety of educational resources and member benefits (including an annual educational conference which attracts 800 attendees from around the globe) the association helps individuals start and maintain successful programs. With more than 55,300 volunteers, nearly 4,700 certified instructors and professionals, 7,672 equines and thousands of contributors from all over the world PATH Intl. inspires and enriches the human spirit.

## PHA – Polish Hippotherapeutic Association

100 Nowoursynowska St.
02-797 Warsaw
POLAND
Contact: Anna Struminska
Phone/Fax: 48 22 59 31 900
Email: biuro@pthip.org.pl
Website: www.pthip.org.pl

The Polish Hippotherapeutic Association (PHA) was founded in 1992 and has been a member of HETI since 1999. The Society was established by doctors, therapists and horsemanship instructors who since the 80's engaged in introducing hippotherapy into the rehabilitation of people with disabilities, especially for children with cerebral palsy. In Poland, the word hippotherapy defines all therapeutic activities associated with horses and riding, therefore PHA offers not only physiotherapy with the horse but also therapeutic riding and even recreation and horsemanship sports for people with disabilities.

The mission of PHA includes:

- supporting and promoting hippotherapy as a rehabilitation method
- supporting and promoting hippotherapy as beneficial within a pedagogic, psychotherapeutic and sociotherapeutic context
- improving and unifying hippotherapists' skills
- providing members access to professional literature and information about hippotherapy's progress in Poland and abroad

The Association promotes the best hippotherapeutic centers by supporting them with their patronage and with individual organization or by lending patronage to events of hippotherapeutic organizations, such as conferences and training sessions.
Specialists at PHA established a training program for hippotherapists and issued professional qualifications. 1612 people have been trained by that program since its inception in 1994. Experienced therapists from the Association also developed a set of basic rules of conduct for hippotherapists, the Canon of Polish Hippotherapy. Furthermore, PHA publishes its own magazine, Hippotherapeutic Review, to promote professional, safe and responsible hippotherapy.

## PT-CH – Pferdegestützte Therapie Schweiz

Bergstrasse 77
5430 Wettingen
SWITZERLAND
Contact: Luz Sozzi
Phone: +41 (56) 427 14 64
+41 (79) 442 95 58
Email: praesidium@sv-hpr.ch
Website: www.sv-hpr.ch/de/Pferdegestuetzte-Therapie-1

PT-CH Horse-assisted Therapy Switzerland (Pferdegestützte Therapie Schweiz – PT-CH) was founded in 1985 as the Swiss Union for Therapeutic Riding (Schweizerische Vereinigung für Heilpädagogisches Reiten - SV-HPR) to promote among experts and the public the recognition of therapeutic riding. Main tasks of the PT-CH are:

- Coordinating and supporting the activities of its members and representing their interests.
- Offering the training of riding therapists, certified by the PT-CH (approx. 1000 Hours)
- regulating the training of riding therapists with the federal professional recognition as the goal.
- Promote training courses for its members in the field of horse-assisted activities and handling of horses.
- Promoting animal training and breeding of horses for therapeutic use. To that end the PT-CH awards a plaque to quality model stables in recognition of their work in the service of people with disabilities.

PT-CH is a member of the Federation of Horses in Education and Therapy International (HETI).

## Qi Zhi Vocational Training Center (The Therapeutic Riding Centre of Taiwan)

77 Yu-Ying Road
Zhongli, Taoyuan County
32057 Taiwan
Contact: Wanda Wang
Phone: +8863/4627912
Email: thwang@thrct.org.tw
Website: www.thrct.org.tw

Qi Zhi Vocational Training Center has been founded in 1977 and is providing community-based residential services and employment services for mentally challenged people.

The Therapeutic Riding Centre of Taiwan (ThRCT) was established under the organization in 2004. It provides Equine Assisted Education and Rehabilitation for physically and mentally challenged persons of all ages. The program includes individual classes, group classes, outings for resource classes, riding camps, training of competitive riders, competitions for mentally and physically challenged riders and international exchange. Three of Qi Zhi's service receivers are currently working at the stable.

A first group of instructors was certified by HETI in 2013. There are also twoAHA trained physiotherapists working at the centre.

Every second year the annual "Hope Cup" is held as an International Competition for Physically and Mentally Challenged Novice Riders from the Asian-Pacific Region. The next international Hope Cup is planned for 2017.

Qi Zhi was part of the organization team of the 2015 HETI Congress in Taiwan. The hard work of its staff and volunteers were the key to the success of the Congress.

HONG KONG

駎騁

# RDA – Riding for the Disabled Association Hongkong

Lot 45
Lung Mun Rd.
Tuen Mun, N.T.
HONG KONG SAR
Contact: Apple Wong
Phone: (852) 2454 9961
Fax: (852) 2465 5914,
Email: info@rda.org.hk
Website: www.rda.org.hk

Riding for the Disabled Association Ltd. (RDA) was founded in Hong Kong in 1975 by the then Governor's wife, Lady MacLehose. A non-profit organization, RDA provides riding instruction and other associated activities to adults and children with disabilities. Their aim is to teach these members of their community to ride a horse to the best of their ability and to learn to interact with the horse and each other. Currently, RDA has 20 horses and ponies. Each year, it offers 28 free weekly sessions for 36 weeks at 3 RDA Centers in Pokfulam, Tuen Mun and Lo Wu. Each session lasts for 45 minutes to an hour and accommodates 6 riders. In all, RDA offers in excess of 6,000 lessons per year. In addition there are lessons run for larger groups so that schools and other organizations can decide if riding would be beneficial.

Competitive riding is a more recent addition to RDA's offerings. Late in 2005, the Sports Sub Committee of the RDA General Committee was formed with a mission to find and train Para Equestrians for competition in local, regional, international and the Paralympics. Currently, RDA has identified a few young athletes who have potential and provides regular training to these young athletes, working towards having them compete in local shows and hopefully international competitions in the future.

# RDA – Riding for the Disabled Association Ireland

97 Sandyford Downs
Sandyford Village,
Dublin D18 R2F9
IRELAND
Contact: Dilys Lindsay
Phone: 353 86 8146898
Email: rdaisecretary@eircom.net
Website: www.rdai.org

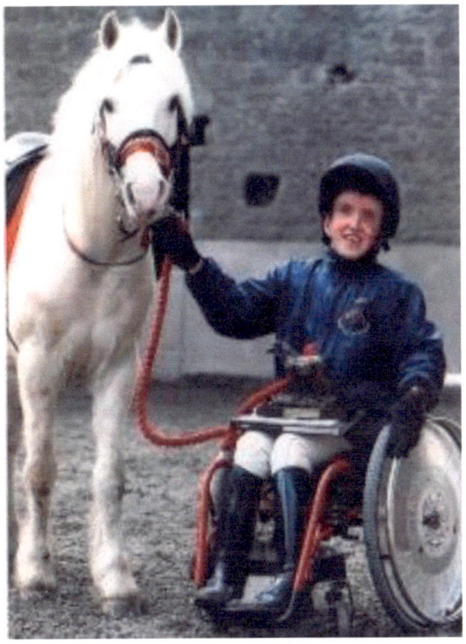

RDA Ireland, founded in 1969, is a voluntary organisation providing opportunities for riding and carriage driving to people with intellectual or physical disabilities. The main focus of RDA Ireland is to broaden the physical, intellectual and social horizons of people with special needs.

At present, more than 55 voluntary groups nationally provide weekly lessons to over 700 disabled people. There are classes for riders and drivers of all ages, abilities and disabilities. Funds to run the groups are raised at the national and local level. Groups meet in local equestrian centres with good wheelchair access and horses suitable for teaching a sport on horseback to a disabled person. Each group has its own committee, organiser, riding instructor and therapist (where available) and a number of volunteers and RDAI values the tremendous voluntary commitment of over 600 people. The Council of RDAI coordinates all this activity, arranges insurance, fundraising and publicity. The Training Committee runs local and national Training Days for helpers, instructors and therapists.

Carriage drivers compete locally and at national events in Punchestown and Dublin International Horse Show. They also travel to compete internationally, including European and World Championships. The highlight of the year for our riders is the RDS in August – venue for the Dublin International Horse Show, where eight riders are chosen from around the country to take part in a trail ride, watched by their families, friends and group members. On that day, they are just as important as every other rider who qualifies to compete in the Dublin International Horse Show. Every year RDAI hold their National AGM and Conference at a different location around the country. Volunteers and some riders attend this event, at which awards are presented to the Rider of the Year and the Volunteer of the Year, as well as long service awards for helpers and ponies/horses.

## RDA – Riding for the Disabled Association Japan

2-21-1 Utsukushigaoka nishi,
Aoba-ku, Yokohama City,
Kanagawa 225-0001
JAPAN
Contact: Ms. Kiginu Nakata
Phone: +81 80 6519 7818
Email: rda@rdajapan.com
Website: www.rdajapan.com

RDA Japan is a certified non-profit organization which promotes and supports safe riding and activities with the horse for people with disabilities in order to improve their quality of life. For this purpose, they

- organize workshops and seminars on therapeutic riding and equine issues, as well as on welfare of horses
- offer training courses for instructors, volunteer helpers and therapists
- support unit-groups to improve their activities
- assist new groups in their start-up phase
- send qualified instructors to RDA-related activities
- provide support to groups through our network of therapists
- engage in international exchange
- promote awareness of the general public on equine-assisted activities

At the 2010 unit meeting, the members will come together to share their experience and discuss topics of common interests. This year's main theme is safety. For the second half of the year the Instructor's Certificate Test is being planned and RDA looks forward to welcoming new instructors. Another important task this year is to strengthen the ties with other organizations and cooperate in promoting therapeutic riding in our country.

## RDA – Riding for the Disabled Association UK

Norfolk House
1 Tournament Court
Edgehill Drive
Warwick CV34 6LG
UNITED KINGDOM
Contact: Mrs. Julie Dargie
Phone: +44 (0)845 658 1082
Email: info@rda.org.uk
Website: www.rda.org.uk

RDA has spent 40 years delivering opportunities for therapy, achievement and enjoyment to people with disabilities. Founded as a national organisation in 1969, RDA now has 500 member groups across the UK – together, these groups help 28,000 people to take part in riding, carriage driving and vaulting activities.

Although set up to work with people with physical disabilities, RDA Groups now work across the spectrum of disabilities, as well as age range, social status and urban and rural environments. The focus of RDA's work is to ensure that each individual derives a direct therapeutic benefit and has the opportunity to achieve their personal goal (whether that be learning a very simple skill or winning a major competition).

RDA gives clients the opportunity to:

- Reach therapeutic goals
- Achieve their personal ambitions
- Combat social isolation
- Develop life skills
- Experience the outdoors
- Connect with animals

Although the current level of service delivery is impressive, there are many more people who merit help and support. Currently, RDA does not fully meet the increasing demand as the vast majority of RDA Groups are reporting a waiting list. RDA's focus, therefore, is on increasing capacity to serve even more clients.

## RDAA – Riding for the Disabled Association of Australia Ltd

PO Box 310
Torrensville SA 5031
AUSTRALIA
Contact: Theresa Moore
Phone: 61 (0)8 8354 4555
Fax: 61 (0)8 8354 4777
Email: admin@rda.org.au
Website: www.rda.org.au

Riding for the Disabled Association Australia (RDAA) is a nonprofit organisation which provides opportunities for anyone with a disability to enjoy safe, healthy, stimulating, therapeutic, horse-related activities. RDAA develops ability and enriches lives through specialised programs and relationships with horses. RDAA:

- is the federal body of 7 autonomous RDA member states and territories in Australia;
- acts as a national body and resource for its member states and territories to access accreditation, education and insurance;
- encourages national links between the 7 affiliated states and territories offering therapeutic riding, driving and equine related activities;
- is actively involved in establishing national policies and standards to ensure excellence in service delivery of RDA programs across the country;
- liaises with government and industry on national issues;
- leads in equine matters related to therapeutic riding;
- plays a significant role as a member of the Federation of Horses in Education and Therapy International (HETI)
- promotes elite athletes at all levels

From its mission statement:
"RDA enables people with disabilities throughout Australia to experience enjoyment, challenge and a sense of achievement through participation in equestrian activities to improve their quality of life, attain personal goals and develop life skills."

## SARDA – South African Riding for the Disabled

P.O. Box 235
Constantia, 7848, Western Cape
SOUTH AFRICA
Contact: Belinda Thom-Bain
Phone: 27 82 449 5176
Email: national@sarda.co.za
Website: www.sarda.co.za

South Africa is a beautiful and diverse country; 1,2 million square kilometres in size with a population of about fifty million. Well known for its sunny climate, magnificent scenery and of course Madiba (Nelson Mandela), this year South Africa was the first ever African nation to host the soccer world cup.

South Africa's population is multicultural and multiracial and the 400 plus riders that visit SARDA's six facilities each week are drawn from all groups. To ensure that riding is available to all there is no charge for any of the sport riding and riding skills classes offered. Despite a rigorous instructor qualification process SARDA's twenty-five instructors, along with many helpers, work on a voluntary basis. SARDA employs only eight staff members.

Development of PARA riders falls under SARDA's umbrella as well and in November 2009, the Cape Town branch hosted the first ever FEI 3* event held in South Africa. Thanks to a generously awarded scholarship, SARDA is now also instrumental in the setting up and running of an Equine Assisted Activities & Hippotherapy Association.

As a Third World country South Africa has a small tax base so no government financial support is available. SARDA is a registered non-profit organisation and keep their riders in the saddle thanks to fund raising efforts, generosity of the public and sympathetic donors.

## SRT – Finnish Association of Equine Facilitated Therapies

Välitalontie 11
87850 Paltaniemi
FINLAND
Contact: Sanna Mattila-Rautiainen
Phone: 358 41 454 2428
Email: sanna.mattila@saunalahti.fi
Website: www.suomenratsastusterapeutit.net

The Finnish Association of Equine Facilitated Therapies - SRT was founded in 2000. It serves as a link between its members, attends to their interests and the interests pertaining to the field of Equine Facilitated Therapies in general. One of SRT's goals is to help its members to continually improve their professional skills and knowledge through courses and lectures. The association provides guidelines to the quality, ethics and practice in Finland. They have created an active national and international cooperation with other organizations that share related interests.

A training program started in Finland in 1988. The training is completed in approximately three years. The entrance requirements to the training are a college degree in a health care profession, pedagogy or social sciences and two years work experience in the field of rehabilitation after graduation. Good horsemanship and riding skills are required. As part of the training program in 2013 a new course starts in Hevosopisto Equine College, carried out with University of Turku and supervised by the Association of Finnish Riding Therapists.

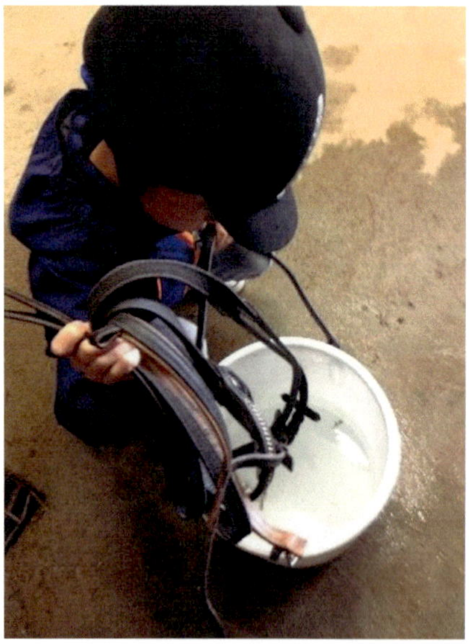

SRT has also created an active national and international cooperation with other organizations that share related interests in therapeutic riding. The Association of Finnish Riding Therapists has been a full member of HETI since 2002. 2010-2012 it was a silent partner to EU funded Hippotherapy- Shared Learning Opportunities created between Hevosopisto Equine College with ACPTR U.K. and the Polish Hippotherapeutic Association.

## TRAG – Therapeutic Riding Association of Greece

P. Kanellopoulou 10
Goudi
11527 Athens
GREECE
Contact: Dimitra Karouzaki
Phone/Fax: 0030 210 7488566
Email: triding@otenet.gr
Website: www.trag.gr

The Therapeutic Riding Association of Greece (T.R.A.G.) was founded in 1992 as a voluntary and nongovernmental organization. A year later, it acquired its athletic identity. In 2004, TRAG obtained its own center of riding for people with specials needs in Goudi, Athens. Its board of directors consists of 7 members who all are volunteers. It is elected every 4 years after the general assembly and is voted on by the members of the Association.

The workforce of the Association consists of seven people and nine horses and the Association offers services to over 150 clients weekly. This wouldn't be possible without the help of more than 70 volunteers (Greek and EVS volunteers from different countries) and of course many known and unknown sponsors who have been supporting the Association from its very beginning.

Within the bounds of increasing progress and improvement, Therapeutic Riding Association of Greece introduces its collaboration with the German Association of Therapeutic Riding, Deutsches Kuratorium für Therapeutisches Reiten e.V. (DKThR), for certification as Specialist for psycho-educational therapy with horses (DKThR/TRAG). Everyone can find more information in www.trag.gr or contact us at triding@otenet.gr or 00302107488566. Our mission is to offer quality and safe Therapeutic Riding sessions to as many people with disabilities as possible and also train therapeutic riding instructors in order to expand Therapeutic Riding as a method of rehabilitation all over Greece.

## Turkish Equestrian Federation

Poligon Caddesi No 4
Sarıyer / Istanbul 3446
TURKEY
Contact: Ms. Yasemin Inan
Tel: 90 532 277 2057
Email: yaseminan@hotmail.com
Website: www.binicilik.org

The Turkish Equestrian Federation has a PATH Intl. certified instructor giving adaptive riding lessons in one of their clubs. Under the project name 'My horse,my wing', they are working hard to apply the regulations and rules of adaptive riding and hippotherapy based on international standarts to raise new riders for the paraolympics.

They have signed a protocol for the project with one of their sports clubs on the 3rd of December, the International Day of People with Disabilities. The name of this sports club is Galatasaray and the website of this club is www.galatasaray.org. Galatasaray is particularly important for the federation because the club is their first sponsor for the project. The club is particularly well known for its football team, but as a large club it is dealing with many areas of sports including riding. To that end, the president of the club had his picture taken on that day with the children and horses to demonstrate the importance of adaptive riding and hippotheraphy.

## WITH – Welsh Institute of Therapeutic Horsemanship

Pen Y Trip Farm
Treflys
Porthmadog
Gwynedd
LL49 9YL
UNITED KINGDOM
Contact: Iain Crosbie, Director of Operations
Phone: +44 (0)1766 512 026
Email: info@therapeutichorsemanship.org
Website: www.therapeutichorsemanship.org

WITH helps disadvantaged young people to develop confidence and communication skills through working on the ground with horses as well as providing training in the field, promoting standards and establishing a network of approved centres. We work with young people aged 8-25 from across North Wales, many of whom face multiple disadvantages and might never normally have the opportunity to spend time around horses. All our services are provided free of charge.

**The Penytrip Project** – Through helping to rehabilitate a rescued horse, young people learn patience, empathy and the value of hard work, gaining a set of practical and 'soft' skills that will enhance their future prospects.

**The Starfish Project** – tackles the issue of truancy, low school attendance and exclusion from education using Therapeutic Horsemanship sessions and outdoor education.

**Saturday Club** – Participants learn how to train and care for horses, take part in Forest School activities, horse agility competitions, crafts, quizzes and treasure hunts. It is a great way for young people to meet new like-minded people, enjoy a shared interest, socialise and continue to build their confidence and communication skills in a safe, supportive environment.

**Young Volunteers Scheme** – Participants are supported to volunteer at WITH in order to gain inter-personal and practical skills that will enhance their future chances of employment and further education.

**Therapeutic Horsemanship Practitioner Training** – Designed to equip potential practitioners with the basic knowledge and skills that they need to create their own groundwork-based equine programs which promote mental health and emotional well-being.

:

# HETI Membership Directory 2015/16

# Business Members

## Unbridled Therapy, Florida, United States

Equine Facilitated Psychotherapy Services, Training for Mental Health Professionals and for Equine Specialists with Marilyn Sokolof

- *Licensed Psychologist*
- *PATH Certified Therapeutic Riding Instructor*
- *PATH Equine Specialist in Mental Health/Learning*
- *Clinical Director of HorseMpower, Inc.*
- *Past President of Equine Facilitated Mental Health Association*

HORSE-ASSISTED PSYCHOTHERAPY
WITH MARILYN SOKOLOF PHD

**Info: www.unbridledtherapy.com • msokolof@aol.com**

## British Horse Society, United Kingdom

With over 89,000 members, The British Horse Society is the UK's largest and most influential equestrian charity. We protect and promote the interests of all horses and those who care about them, including the 3.5 million people in the UK who ride or who drive a horse-drawn carriage, by

- *Providing welfare services for horses and advice for owners*
- *Campaigning for and protecting equestrian rights of way*
- *Improving safety for horses, riders and carriage drivers*
- *Promoting riding for all*

**www.bhs.org.uk • 02476 840500 (join, renew or donate) • 02476 840517 (report a horse in distress)**

## Gaits of Change, United States

*Gaits of Change offers a variety of programs, workshops and consulting opportunities, customized to fit your specific needs. Focus is on education, good leadership, best practices and living the mission of the organization.*

- *Program Consulting & Strategic Planning*
- *Registered & Advanced Instructor Training*
- *Mentorship*
- *Staff Development & Motivation*
- *Interactive Vaulting Programs*
- *Horse Training*

**www.gaitsofchange.com • Sandy@GaitsOfChange.com • 949-701-5271**

## High Hopes Therapeutic Riding, Inc. Connecticut, United States

Study at a PATH Intl. Premier Accredited Center. Extensive training and education opportunities for the international community.

- *Courses taught to the PATH Intl. Advanced Instuctor Level*
- *Certification for therapeutic riding instructors at the Registered and Advanced levels*
- *Classes in therapeutic riding center administration and management*
- *More than 200 students from the U.S. and 15 countries trained*

**36 Town Woods Road • Old Lyme, CT 06371 USA • 860-434-1974 • www.highhopestr.org**

## equinoterapiaarg Equinoterapia La Rioja, Argentina

*Equinoterapiaarg offers a wealth of information on equine assisted activities in Argentina. Throughout the whole country – and spread across eleven provinces – there are many centers where the discipline of equine therapy is practiced.*
*Almost 60 centers and their contact information can be accessed through this site.*

**www.equinoterapiaarg.com.ar**

## Sunshine Coast, Queensland, Australia

*Sunshine Coast Riding for the Disabled provides equine-based therapeutic and recreational riding and driving programs for the disability community on the Sunshine Coast, Queensland:*

- *Hippotherapy*
- *EAGALA and EAL programs*
- *recreational riding*
- *carriage driving*
- *para-equestrian training*

**www.sunshinecoastrda.com • admin@sunshinecoastrda.com**

# HETI Membership Directory 2015/16

# Associate, Institute, and Honorary Members

# Associate Members

A.V.R.E.S., ITALY
AAS DISCAN., SPAIN
Abrams, Barbara, USA
Adams-Wheeler, Sue, UNITED KINGDOM
AIASPORT ONLUS, ITALY
Alden, Ann, USA
APRES, ITALY
Arteveldehogeschool, BELGIUM
Asociacionde Equinoterapia de Costa Rica, Costa Rica
Assoc. Reg. D'Equitation Therap. D'Alsace, FRANCE
Association Hippos, SLOVENIA
Athanasopoulou, Spiridoula, GREECE
AWO Dortmund, Henrike Struck, GERMANY
Barbara Engel Therapy Services, USA
Barnfield, Anne, CANADA
Barrett, Louise, UNITED KINGDOM
Baxevanidi, Chrysoula, GREECE
Beitostolen Helsesportsenter, NORWAY
Benavides, Maria Angelica Rodriguez, COLOMBIA
Bender, Renate, Chile
Benjamin, Joann, USA
Boyce (Rufus), Sharon, SOUTH AFRICA
Broersma, Trish, USA
Brophy, Mary R., IRELAND
Brown, Octavia, USA
Bruna Grubits Freire, Heloisa, BRAZIL
Buyse, Brigitte, BELGIUM
Cabre, Geni, UNITED ARAB EMIRATES
Cain, Anja, USA
Caseys Healing Hooves, USA
Cavalo Amigo, PORTUGAL
Centro de Equinoterapia (en Guadalajara), MEXICO
Centro de Rehabilitacion Infantil HOGA, MEXICO
Centro Equestre Equovita, BRAZIL
Centro Mexiquense de Equinoterapia, MEXICO
Chang, Chao-Yuan, REPUBLIC OF CHINA (TAIWAN)
Chen, Chun-Mei, REPUBLIC OF CHINA (TAIWAN)
Chengdu RDA, CHINA
Chomoriti, Ekaterini, GREECE
Claremont Therapeutic Riding Centre, Inc., AUSTRALIA
Cutler, Joan, USA

# Associate Members (continued)

Cohen, Anna, AUSTRALIA
Cunningham, Bonnie, USA
Davey, Daphne, CANADA
De Acuna, Fernanda Morey, PERU
Donahue, Jen, USA
Equest, USA
Equinesport Emotional Management, SPAIN
F.Q.E.T., CANADA
Faubel, Giselle, USA
Field of Dreams Foundation, UNITED KINGDOM
Förderkreis für Therapeutisches Reiten, GERMANY
Fulcrum Therapy LLC, USA
German Research Center for Equine Assisted Therapy, GERMANY
Gkampriella, Lappa, GREECE
Goeteborgs Handikappridklubb, SWEDEN
Goetske, Birgitte, DENMARK
Goldberg, Nicola Rut, ISRAEL
Gomes Pereira, Joana Luisa, PORTUGAL
Grannell, Marian, UNITED KINGDOM
Green Chimneys, USA
Gross-Naschert, Edith, MEXICO
Haartz, Elizabeth, USA
Hakanson, Margareta, SWEDEN
Hipoterapi Antalya, TURKEY
Hippotherapy RDA WA, AUSTRALIA
Hoff, Jean Paul, FRANCE
Hoffmann, Gerlinde, GERMANY
Houry, Judy CTRI, MBE, SAUDI ARABIA
Hutton-Anetopoulou, Sara, GREECE
IAPE EQUO, BRAZIL
Institut für tiergestützte Ausbildung - ITA, GERMANY
Jennings, Helen, MALAYSIA
Kamizi, Kallirroi, GREECE
Kay, Judi, ISRAEL
Kedra, Foteini, GREECE
Kellogg, Lili, USA
Klouman, Marianne, NORWAY
Kreindler, Dalia Assoc., ISRAEL
KRILA- terapijsko jahanje, CROATIA
Kulichova, Jana, CZECH REPUBLIC

# Associate Members (continued)

Kwon, Jeong Yi, KOREA
Lac, Veronica, USA
Lalousi, Despoina, GREECE
Lanark County Therapeutic Riding Program, CANADA
Laurie, Harriet, UNITED KINGDOM
Lercari, Lydia, URUGUAY
Lightsey, Priscilla, USA
Lima, Syllas, BRAZIL
Lipponen, Maija, FINLAND
Liu, Ming, CHINA
Longden, Mary, AUSTRALIA
Love Quest Children's Foundation, USA
Lysell, Jeanette, NORWAY
Macauley, Beth, USA
Mane Stream, USA
Mansfield, Marny, USA
Masako, Nakata, JAPAN
Masatoshi, Nakata, JAPAN
Mastorou, Perdika, GREECE
Matsouki, Eleni, GREECE
Mattila-Rautiainen, Sanna, FINLAND
McIntyre Centre, AUSTRALIA
Mehlem, Monika, GERMANY
Ming-Hui, Chang, REPUBLIC OF CHINA (TAIWAN)
Mitropoulou, Anna, GREECE
Morris, Teresa, USA
Municipeo de San Nicolas, MEXICO
Nakai, Yukari, JAPAN
Northern Lights TR Association, CANADA
Olson, Alissa, USA
O'Shallie, Ann, USA
Pal-O-Mine Equestrian Inc., USA
Panagou, Nantina, GREECE
Pantazelou, Panagiota, GREECE
Papathanasiou, Styliani, GREECE
Pearce, Norma, AUSTRALIA
Pegasus RDA of the ACT, AUSTRALIA
Pei-Ying, Lee, REPUBLIC OF CHINA (TAIWAN)
Perry O.A.M., Pauline, AUSTRALIA
Petriki, Petroula, GREECE

# Associate Members (continued)

Pisina, Aspasia, GREECE
Plaza, Alison, USA
Promitheas A.M.K.E., GREECE
RD Samsung, SOUTH AFRICA
RDA (NSW), AUSTRALIA
RDA Australia Hippotherapy, AUSTRALIA
RDA SA Inc., AUSTRALIA
Renker, Lorrie, USA
Richter, Tatjana, GREECE
Rindfleisch-Wu, Uta, REPUBLIC OF CHINA (TAIWAN)
Robertson, Jean Marie, IRELAND
Rosas Rodriguez, Jeannette, COLOMBIA
Sands O.A.M., Ann, AUSTRALIA
SARDA Gauteng, SOUTH AFRICA
Sayler, Pat, USA
Schoerle, Armgard, GERMANY
Schulz, Marietta, GERMANY
Shkedi, Anita, ISRAEL
Shu-Ya, Chan, REPUBLIC OF CHINA (TAIWAN)
Silkwood-Sherer, Debbie, USA
Sistema Municipal DIF de Puebla, MEXICO
SOD ITALIA, ITALY
Solana, Laura, MEXICO
Sosiaaliped. Hevostoimintayhdistys, FINLAND
Spirit Open Equestrian Program, Inc., USA
Splinter-Watkins, Kathryn, USA
Sponsel, Lelia, CANADA
St. Andrews Presbyterian College, USA
Strides to Success, USA
Sugarman, Peggy, USA
Sweeney, Molly, USA
The Tiltas Trust, UNITED KINGDOM
Thelle, Mona, NORWAY
Therapeutic Riding Centre of Israel, ISRAEL
Touchstone Farm, USA
Tourlakis, Vasilis, GREECE
Triantafyllidou, Georgia, GREECE
Triantouli, Eleni, GREECE
Tsai, Yu-Ling, REPUBLIC OF CHINA (TAIWAN)
Tsung-Hsin, Weng, REPUBLIC OF CHINA (TAIWAN)

# Associate Members (continued)

Tuck, Susanne, AUSTRALIA
Tzalali, Christina, GREECE
Tze-Hsuan, Wang, REPUBLIC OF CHINA (TAIWAN)
UNH Therapeutic Riding, USA
Vidali, Eleni, GREECE
Viruega, Helene, FRANCE
Vitoula, Olga, GREECE
Wagner, Cynthia, USA
Walker, Carolyn, UNITED KINGDOM
Wang, David, REPUBLIC OF CHINA (TAIWAN)
Wheelwright, Christine, USA
Wilson College, USA
Ya-Hui, Li, REPUBLIC OF CHINA (TAIWAN)
Yi-Hung, Chang, REPUBLIC OF CHINA (TAIWAN)
Yorke, Jan, CANADA
Yu-Jen, Sun, REPUBLIC OF CHINA (TAIWAN)
Yu-Shan, Lu, REPUBLIC OF CHINA (TAIWAN)
Zepeda, Francisco, MEXICO
Zografos, Vasileios, GREECE

# Institute Members

Able the Disabled Foundation, THAILAND
Carlisle Academy, USA
Colorado State University, USA
Corpoalegria, COLUMBIA

# Honorary Members

Ernst Reinhardt Verlag, GERMANY
Horses and Human Research Foundation, USA
IAHAIO, INTERNATIONAL ORGANIZATION

# HETI Membership Directory 2015/16

# Index of Members by Country

# Index of Members by Country


## Argentina

Equinoterapia La Rioja (Business Member)

## Australia

Claremont Therapeutic Riding Centre, Inc. (Associate Member)
Cohen, Anna (Associate Member)
Hippotherapy RDA WA (Associate Member)
Longden, Mary (Associate Member)
McIntyre Centre (Associate Member)
Pearce, Norma (Associate Member)
Pegasus RDA of the ACT (Associate Member)
Perry O.A.M., Pauline (Associate Member)
RDA (NSW) (Associate Member)
RDA Australia Hippotherapy (Associate Member)
RDA SA Inc. (Associate Member)
RDAA – Riding for the Disabled Association of Australia Ltd (Federation Member)
Sands O.A.M., Ann (Associate Member)
Sunshine Coast Riding (Business Member)
Tuck, Susanne (Associate Member)

## Austria

Österreichische Initiative – Pferde helfen Menschen in cooperation with e.motion (Federation Member)

## Belgium

Arteveldehogeschool (Associate Member)
Buyse, Brigitte (Associate Member)
HANDHIPPO - Belgische Nationale Vereniging van Therapeuten Gespecialiseerd in Hippotherapie / Union Nationale Belge des Thérapeutes Spécialisés en Hippothérapie (Federation Member)

## Brazil

ANDE-BRASIL – Associação Nacional de Equoterapia (Federation Member)
Bruna Grubits Freire, Heloisa (Associate Member)
Centro Equestre Equovita (Associate Member)
IAPE EQUO (Associate Member)
Lima, Syllas (Associate Member)

## Canada

Barnfield, Anne (Associate Member)
CanTRA – Canadian Therapeutic Riding Association (Federation Member)
Davey, Daphne (Associate Member)
F.Q.E.T. (Associate Member)
Lanark County Therapeutic Riding Program (Associate Member)
Northern Lights TR Association (Associate Member)
Sponsel, Lelia (Associate Member)
Yorke, Jan (Associate Member)
KRILA- terapijsko jahanje (Associate Member)

# Index of Members by Country (continued)


## Chile

Bender, Renate (Associate Member)

## Colombia

Benavides, Maria Angelica Rodriguez (Associate Member)
Rosas Rodriguez, Jeannette (Associate Member)
Corpoalegria (Institute Member)

## Costa Rica

Asociacionde Equinoterapia de Costa Rica (Associate Member)

## Croatia

KRILA- terapijsko jahanje (Associate Member)

## Czech Republic

Kulichova, Jana (Associate Member)
CHS – Czech Therapeutic Riding Association (Federation Member)

## Denmark

Danish Ridefysioterapi (Federation Member)
Goetske, Birgitte (Associate Member)

## Finland

Lipponen, Maija (Associate Member)

## Finland (cont'd)

Mattila-Rautiainen, Sanna (Associate Member)
Sosiaaliped. Hevostoimintayhdistys (Associate Member)
SRT – Association of Finnish Riding Therapists (Federation Member)

## France

Assoc. Reg. D'Equitation Therap. D'Alsace (Associate Member)
Fédération Nationale Handi cheval (Federation Member)
FENTAC – Fédération Nationale de Thérapie avec le Cheval (Federation Member)
Hoff, Jean Paul (Associate Member)
Viruega, Helene (Associate Member)

## Germany

AWO Dortmund, Henrike Struck (Associate Member)
DKThR – Deutsches Kuratorium für Therapeutisches Reiten E.V. (Federation Member)
Ernst Reinhardt Verlag (Honorary Member)
FN – Fédération Equestre Nationale / Deutsche Reiterliche Vereinigung e.V (Federation Member)
Förderkreis für Therapeutisches Reiten (Associate Member)
German Research Center for Equine Assisted Therapy (Associate Member)
Hoffmann, Gerlinde (Associate Member)

# Index of Members by Country (continued)

## Germany (cont'd)

Institut für tiergestützte Ausbildung - ITA (Associate Member)
Mehlem, Monika (Associate Member)
Schoerle, Armgard (Associate Member)
Schulz, Marietta (Associate Member)

## Greece

Athanasopoulou, Spiridoula, (Associate Member)
Baxevanidi, Chrysoula, (Associate Member)
Chomoriti, Ekaterini, (Associate Member)
ELETHIP – Hellenic Association for Therapeutic Riding, (Federation Member)
Gkampriella, Lappa, (Associate Member)
HETRICS/KETHIS – Hellenic Therapeutic Riding Center of Serres, (Federation Member)
Hutton-Anetopoulou, Sara, (Associate Member)
Ippostirixi – Therapeutic Riding, Hippotherapy and Education, (Federation Member)
Kamizi, Kallirroi, (Associate Member)
Kedra, Foteini, (Associate Member)
Lalousi, Despoina, (Associate Member)
Mastorou, Perdika, (Associate Member)
Matsouki, Eleni, (Associate Member)
Mitropoulou, Anna, (Associate Member)
Panagou, Nantina, (Associate Member)
Pantazelou, Panagiota, (Associate Member)
Papathanasiou, Styliani, (Associate Member)
Petriki, Petroula, (Associate Member)

## Greece (cont'd)

Pisina, Aspasia, (Associate Member)
Promitheas A.M.K.E., (Associate Member)
Richter, Tatjana, (Associate Member)
Tourlakis, Vasilis, (Associate Member)
TRAG – Therapeutic Riding Association of Greece, (Federation Member)
Triantafyllidou, Georgia, (Associate Member)
Triantouli, Eleni, (Associate Member)
Tzalali, Christina, (Associate Member)
Vidali, Eleni, (Associate Member)
Vitoula, Olga, (Associate Member)
Zografos, Vasileios, (Associate Member)

## Hong Kong

RDA - Riding for the Disabled Association Hongkong (Federation Member)

## Hungary

Hungarian Riding for the Disabled Federation (Federation Member)

## Ireland

Brophy, Mary R. (Associate Member)
Festina Lente (Federation Member)
RDA – Riding for the Disabled Association Ireland (Federation Member)
Robertson, Jean Marie (Associate Member)

# Index of Members by Country (continued)


## Israel

Goldberg, Nicola Rut (Associate Member)
Kay, Judi (Associate Member)
Kreindler, Dalia Assoc. (Associate Member)
Shkedi, Anita (Associate Member)
Therapeutic Riding Centre of Israel (Associate Member)

## Italy

A.V.R.E.S. (Associate Member)
AIASPORT ONLUS (Associate Member)
ANIRE – Italian National Association of Therapeutic Riding (Federation Member)
APRES (Associate Member)
Centro di Riabilitazione Equestre Vittorio di Capua (Federation Member)
CNRAFA ONLUS – Italian National Centre of Research, Therapy and Training (Federation Member)
FISE – Federazione Italiana Sport Equestri / Italian Equestrian Federation (Federation Member)
LAPO – Italian Association of Families, Bodies and Professions against Neurological and Psychiatric Diseases in Childhood and Adolescence (Federation Member)
SOD ITALIA (Associate Member)

## Japan

Horse Community Piskari Japan (Federation Member)
JTRA – Japan Therapeutic Riding Association (Federation Member)
Masako, Nakata (Associate Member)
Masatoshi, Nakata (Associate Member)
Nakai, Yukari (Associate Member)
RDA – Riding for the Disabled Association Japan (Federation Member)

## Korea

Kwon, Jeong Yi (Associate Member)

## Luxembourg

ATE – Association Therapie Equestre (Federation Member)

## Malaysia

Jennings, Helen (Associate Member)

## Mexico

AMET – Mexican Association of Therapeutic Riding (Federation Member)
Centro de Equinoterapia (en Guadalajara) (Associate Member)

# Index of Members by Country (continued)


## Mexico (cont'd)

Centro de Rehabilitacion Infantil HOGA (Associate Member)
Centro Mexiquense de Equinoterapia (Associate Member)
Gross-Naschert, Edith (Associate Member)
Municipeo de San Nicolas (Associate Member)
Sistema Municipal DIF de Puebla (Associate Member)
Solana, Laura (Associate Member)
Zepeda, Francisco (Associate Member))

## Netherlands

FPG – Federatie Paardrijden Gehandicapten (Federation Member)

## New Zealand

NZRDA - New Zealand Riding for the Disabled Association Incorporated (Federation Member

## Norway

Beitostolen Helsesportsenter (Associate Member)
Hest og Helse (Federation Member)
Norway (cont'd)
Klouman, Marianne (Associate Member)
Lysell, Jeanette (Associate Member)

## Norway (cont'd)

Norwegian Physiotherapist Association (Federation Member)
Thelle, Mona (Associate Member)

## People's Republic of China

Chengdu RDA, (Associate Member)
Liu, Ming, (Associate Member)

## Peru

De Acuna, Fernanda Morey (Associate Member)

## Poland

PHA - Polish Hippotherapeutic Association (Federation Member)

## Portugal

Cavalo Amigo (Associate Member)
Gomes Pereira, Joana Luisa (Associate Member)

## Republic of China (Taiwan)

Chang, Chao-Yuan (Associate Member)
Chen, Chun-Mei (Associate Member)
CTEA – Chinese Taipei Equestrian Association (Federation Member)
Ming-Hui, Chang (Associate Member)
Pei-Ying, Lee (Associate Member)

# Index of Members by Country (continued)


## Republic of China (Taiwan) (cont'd)

Qi Zhi Vocational Training Center (The Therapeutic Riding Centre of Taiwan) (Federation Member)
Rindfleisch-Wu, Uta (Associate Member)
Shu-Ya, Chan (Associate Member)
Tsai, Yu-Ling (Associate Member)
Tsung-Hsin, Weng (Associate Member)
Tze-Hsuan, Wang (Associate Member)
Wang, David (Associate Member)
Ya-Hui, Li (Associate Member)
Yi-Hung, Chang (Associate Member)
Yu-Jen, Sun (Associate Member)
Yu-Shan, Lu (Associate Member)

## Russia

National Federation of Therapeutic Riding (Russia) (Federation Member)

## Saudi Arabia

Houry, Judy CTRI, MBE (Associate Member)

## Slovenia

Association Hippos (Associate Member)

## South Africa

Boyce (Rufus), Sharon (Associate Member)

## South Africa (cont'd)

RD Samsung (Associate Member) South Africa (cont'd)
SARDA - South African Riding for the Disabled (Federation Member)
SARDA Gauteng (Associate Member)

## South Korea

KATH – Korean Academy of Therapeutic Riding (RDA - Korea) (Federation Member)
Let's Run (formerly RDA-KRA) (Federation Member)

## Spain

AAS DISCAN. (Associate Member)
AEDEQ – Asociación Española de Equinoterapias (Federation Member)
Equinesport Emotional Management (Associate Member)
FCA – Fundacion Caballo Amigo (Federation Member)

## Sweden

Goeteborgs Handikappridklubb (Associate Member)
Hakanson, Margareta (Associate Member)
OHI – Organisationen för Hästunderstödda Insatser (Federation Member)

# Index of Members by Country (continued)


## Switzerland

PT-CH – Pferdegestützte Therapie Schweiz (Federation Member)

## Thailand

Able the Disabled Foundation (Institute Member)

## Turkey

Hipoterapi Antalya (Associate Member)
Turkish Equestrian Federation (Federation Member)

## United Arab Emirates

Cabre, Geni (Associate Member)

## United Kingdom

ACPTR – Association of Chartered Physiotherapists in Therapeutic Riding (Federation Member)
Adams-Wheeler, Sue (Associate Member)
Barrett, Louise (Associate Member)
British Horse Society (Business Member)
FCRT – Fortune Centre of Riding Therapy (Federation Member)
Field of Dreams Foundation (Associate Member)
Grannell, Marian (Associate Member)
GUL – God Unlimited Outdoor Therapy (Federation Member)
Laurie, Harriet (Associate Member)

## United Kingdom (cont'd)

RDA – Riding for the Disabled Association UK (Federation Member)
The Tiltas Trust (Associate Member)
Walker, Carolyn (Associate Member)
WITH – Welsh Institute of Therapeutic Horsemanship (Federation Member))

## Uruguay

Lercari, Lydia (Associate Member)

## USA

Abrams, Barbara (Associate Member)
AHA – American Hippotherapy Association (Federation Member)
Alden, Ann (Associate Member)
Barbara Engel Therapy Services (Associate Member)
Benjamin, Joann (Associate Member)
Broersma, Trish (Associate Member)
Brown, Octavia (Associate Member)
Cain, Anja (Associate Member)
Carlisle Academy (Institute Member)
Caseys Healing Hooves (Associate Member)
Colorado State University (Institute Member)
Cunningham, Bonnie (Associate Member)
Cutler, Joan (Associate Member)
Donahue, Jen (Associate Member)
Equest (Associate Member)
Faubel, Giselle (Associate Member)
Fulcrum Therapy LLC (Associate Member)

# Index of Members by Country (continued)

## USA (cont'd)

Gaits of Change (Business Member)
Green Chimneys (Associate MemberHaartz, Elizabeth (Associate Member)
High Hopes Therapeutic Riding (Business Member)
Horses and Human Research Foundation (Honorary Member)
Kellogg, Lili (Associate Member)
Lac, Veronica (Associate Member)
Lightsey, Priscilla (Associate Member)
Love Quest Children's Foundation (Associate Member)
Macauley, Beth (Associate Member)
Mane Stream (Associate Member)
Mansfield, Marny (Associate Member)
Morris, Teresa (Associate Member)
Olson, Alissa (Associate Member)
O'Shallie, Ann (Associate Member)
PACTH – Pennsylvania Council on Therapeutic Horsemanship (Federation Member)
Pal-O-Mine Equestrian Inc. (Associate Member)
PATH International – Professional Association of Therapeutic Horsemanship International (Federation Member)
Plaza, Alison (Associate Member)
Renker, Lorrie (Associate Member)
Sayler, Pat (Associate Member)
Silkwood-Sherer, Debbie (Associate Member)
Sokolof, Marilyn (Business Member)
Spirit Open Equestrian Program, Inc. (Associate Member)
Splinter-Watkins, Kathryn (Associate Member)

## USA (cont'd)

St. Andrews Presbyterian College (Associate Member)
Strides to Success (Associate Member)
Sugarman, Peggy (Associate Member)
Sweeney, Molly (Associate Member)
Touchstone Farm (Associate Member)
UNH Therapeutic Riding (Associate Member)
Wagner, Cynthia (Associate Member)
Wheelwright, Christine (Associate Member)
Wilson College (Associate Member))

## Worldwide

IAHAIO (Honorary Member)

Made in the USA
Las Vegas, NV
03 November 2026

57423826R00062